The Origins of Cauchy's Rigorous Calculus

The Origins of Cauchy's Rigorous Calculus

Judith V. Grabiner

The MIT Press
Cambridge, Massachusetts, and London, England

This book was set in Monophoto Baskerville by Asco Trade Typesetting Ltd., Hong Kong, and printed and bound by The Alpine Press, Incorporated in the United States of America.

Library of Congress Cataloging in Publication Data

Grabiner, Judith V

The origins of Cauchy's rigorous calculus.

Bibliography: p.

Includes index.

1. Calculus—History. 2. Cauchy, Augustin Louis, baron, 1789–1857. I. Title.

QA303.G74 515′.09 80-28969

ISBN 0-262-07079-0

To my parents, Alfred and Ruth Tofield Victor

Contents

Preface

Augustin-Louis Cauchy gave the first reasonably successful rigorous foundation for the calculus. Beginning with a precise definition of limit, he initiated the nineteenth-century theories of convergence, continuity, derivative, and integral. The clear superiority of Cauchy's work over what had come before, and the apparent break with the past that such superiority implies, combine to raise an urgent historical question: Since no great work arises in a vacuum, what, in the thought of his predecessors, made Cauchy's achievement possible? In this book I attempt to answer this question by giving the intellectual background of Cauchy's accomplishment.

Understanding the way mathematical ideas develop has, besides its intrinsic interest, immediate application to our understanding of mathematics and to the teaching of mathematics. The history of the foundations of the calculus provides the real motivation for the basic ideas, and also helps us to see which ideas were—and thus are—really hard. For the convenience of teachers and students of mathematics, I have, whenever possible, cited the mathematics of the past in readily accessible editions. In the appendix I provide English translations of some of Cauchy's major contributions to the foundations of the calculus.

It is a pleasure to acknowledge the many sources of encouragement and support I received while preparing this study. The nucleus of the research was funded by a Fellowship from the American Council of Learned Societies. My PhD research on Joseph-Louis Lagrange, on which part of this study is based, was supported by a doctoral fellowship from the National Science Foundation, and some of the research for the final revision of this manuscript was supported by the National Science Foundation under Grant No. SOC 7907844.

I would like to thank the Harvard College Library for the excellence of its collection and the assistance of its staff. Dirk J. Struik has often given me the benefit of his vast knowledge of eighteenth-century mathematics and has reminded me that the mathematician is a social being as well as a creator of mathematical ideas. Uta Merzbach systematically introduced me to many aspects of the history of mathematics, and first suggested to me that eighteenth-century approximation techniques were worth investigating. Joseph W. Dauben, my colleague at *Historia Mathematica*, gave me the benefit of his criticism of an early version of this book and was generally informative and encouraging. Also, several anonymous referees have provided useful comments at various stages of the preparation of the manuscript. I am grateful to the students in my classes on the history of mathematics at Harvard University, California State University at Los Angeles, Pomona College, and California State University, Dominguez Hills; they have provided both incisive questions and reflective comments. I owe significant intellectual debts to the late Carl Boyer for his *History of the Calculus*, and to Thomas S. Kuhn for his *The Structure of Scientific Revolutions*. Above all I would like to record my gratitude to I. Bernard Cohen of Harvard University. He has been an unfailing source of inspiration as well as of material assistance, and, most important, he taught me what it means to think like a historian. Finally, I would like to thank my husband Sandy Grabiner, who interrupted his own work to spend many hours reading various drafts of this book; his encouragement and mathematical insight have provided the necessary and sufficient conditions for its completion; and my son David, who helped proofread.

Abbreviations of Titles

B. Bolzano, *Rein analytischer Beweis* = *Rein analytischer Beweis des Lehrsatzes dass zwischen je zwey Werthen, die ein entgegengesetztes Resultat gewaehren, wenigstens eine reele Wurzel der Gleichung liege*, Prague, 1817.

A.-L. Cauchy, *Calcul infinitésimal* = *Résume des leçons données a l'ecole royale polytechnique sur le calcul infinitésimal*, vol. 1 [all published], Paris, 1823. In the edition of Cauchy's *Oeuvres*, series 2, vol. 4, pp. 5–261; all references will be to this edition.

A.-L. Cauchy, *Cours d'analyse* = *Cours d'analyse de l'ecole royale polytechnique. 1re partie: analyse algébrique* [all published], Paris, 1821. In the edition of Cauchy's *Oeuvres*, series 2, vol. 3; all page references will be to this edition.

J.-L. Lagrange, *Calcul des fonctions* = *Leçons sur le calcul des fonctions*, new ed., Paris, 1806. In the edition of Lagrange's *Oeuvres*, vol. 10.

J.-L. Lagrange, *Equations numériques* = *Traité de la résolution des équations numériques de tous les degrés*, 2nd ed., Paris, 1808. In the edition of Lagrange's *Oeuvres*, vol. 8.

J.-L. Lagrange, *Fonctions analytiques* = *Théorie des fonctions analytiques, contenant les principes du calcul différentiel, degagés de toute considération d'infiniment petits, d'évanouissans, de limites et de fluxions, et réduits à l'analyse algébrique des quantités finies*, new ed., Paris, 1813. In the edition of Lagrange's *Oeuvres*, vol. 9.

The Origins of Cauchy's Rigorous Calculus

Introduction

What is the calculus? When we ask modern mathematicians, we get two different answers. One answer is, of course, that the calculus is the branch of mathematics which studies the relationships between functions, their derivatives, and their integrals. Its most important subject matter is its applications: to tangents, areas, volumes, arc lengths, speeds, and distances. The calculus can be fruitfully viewed and effectively taught as a set of intuitively understood problem-solving techniques, widely applicable to geometry and to physical systems. Through the generality of its basic concepts and through the heuristic qualities of its notation, the calculus demonstrates the power of mathematics to state and solve problems pertaining to every aspect of science.

The calculus is something else as well, however: a set of theorems, based on precise definitions, about limits, continuity, series, derivatives, and integrals. The calculus may seem to be about speeds and distances, but its logical basis lies in an entirely different subject—the algebra of inequalities. The relationship between the uses of the calculus and the justification of the calculus is anything but obvious. A student who asked what *speed* meant and was answered in delta-epsilon terms might be forgiven for responding with shock, "How did anybody ever think of such an answer?"

These two different aspects—use and justification—of the calculus, simultaneously coexisting in the modern subject, are in fact the legacies of two different historical periods: the eighteenth and the nineteenth centuries. In the eighteenth century, analysts were engaged in exciting and fruitful discoveries about curves, infinite processes, and physical systems. The names we attach to important results in the calculus—Bernoulli's numbers, L'Hôpital's rule, Taylor's series, Euler's gamma function, the Lagrange remainder, the Laplace transform—attest to the mathematical discoveries of eighteenth-century analysts. Though not indifferent to rigor, these researchers spent most of their effort developing and applying powerful methods, some of which they could not justify, to solve problems; they did not emphasize the mathematical importance of the foundations of the calculus and did not really see foundations as an important area of mathematical endeavor.

By contrast, a major task for nineteenth-century analysts like Cauchy, Abel, Bolzano, and Weierstrass was to give rigorous definitions of the basic concepts and, even more important, rigorous proofs of the results of the calculus. Their proofs made precise the conditions under which the relations between the concepts of the calculus held. Indeed, nineteenth-century precision made possible the discovery and application of concepts like those of uniform convergence, uniform continuity, summability, and asymptotic expansions, which could neither be studied nor even expressed in the conceptual framework of eighteenth-century mathematics. The very names we use for some basic ideas in analysis reflect the achievements of nineteenth-century mathematicians in the foundations of analysis: Abel's convergence theorem, the Cauchy criterion, the Riemann integral, the Bolzano–Weierstrass theorem, the Dedekind cut. And the symbols of nineteenth-century rigor—the ubiquitous delta and epsilon—first appear in their accustomed logical roles in Cauchy's lectures on the calculus in 1823.

Of course nineteenth-century analysis owed much to eighteenth-century analysis. But the nineteenth-century foundations of the calculus cannot be said to have grown naturally or automatically out of earlier views. Mathematics may often grow smoothly by the addition of methods, but it did not do so in this case. The conceptual difference between the eighteenth-century way of looking at and doing the calculus and nineteenth-century views was simply too great. It is this difference which justifies our claim that the change was a true scientific revolution and which motivates the present inquiry into the causes of that change.

The most important figure in the initiation of rigorous analysis was Augustin-Louis Cauchy. It was, above all, Cauchy's lectures at the Ecole Polytechnique in Paris in the 1820s that established a new attitude toward rigor and developed many characteristic nineteenth-century concepts and methods of proof. Other mathematicians, of course, participated in the development of the foundations of the calculus. For instance, Karl Friedrich Gauss had an understanding of convergence which, had he treated the subject comprehensively, might have led him to equal Cauchy's accomplishment. Bernhard Bolzano had many ideas, especially about convergence and continuity, which, if more widely disseminated, would have hastened the rigorization of analysis. And, after Cauchy had initiated the new rigorous calculus, many mathematicians—including Cauchy himself—made further contributions in the same spirit. Niels Henrik Abel, schooled in Cauchy's methods, used them to extend the theory of convergence. Bernhard Riemann clarified and extended the concept of the integral. Karl Weierstrass in many ways finished the task Cauchy had begun by systematizing delta-epsilon methods, by emphasizing the distinctions between convergence and uniform convergence and between continuity and uniform continuity, and by eliminating most of the gaps in earlier reasoning. Weierstrass, Eduard Heine, Richard Dedekind, Charles Méray, and Georg Cantor developed the modern theory of real numbers. But all these accomplishments are based on the revolution begun earlier, and in important ways on the work of Cauchy, who created the mathematical climate which made them possible. Thus I shall focus primarily on explaining the origins of Cauchy's achievements.

In looking for the eighteenth-century origins of Cauchy's foundations of the calculus, I shall demonstrate that they grew, not principally out of that period's discussions of foundations but from other, quite different parts of its work. The men who created the major results of eighteenth-century analysis—Newton, Maclaurin, Euler, d'Alembert, Lagrange—unwittingly also developed many of the ideas and methods which later were used to make it rigorous. The work of two of these men is especially noteworthy. Leonhard Euler, whose work historians sometimes take to illustrate the lack of rigor in eighteenth-century calculus, nevertheless developed many techniques that Cauchy would later adapt. Even more important was Joseph-Louis Lagrange. Lagrange was the crucial transitional figure between the eighteenth- and nineteenth-century points of view, in terms of both technique and attitude. He envisioned, though he did not successfully complete, a program to make the calculus rigorous by reducing it to algebra. Furthermore, a number of the techniques Lagrange used in his unsuccessful attempt were saved and effectively employed by Cauchy himself.

Chapter 1 will describe more fully the nature of the nineteenth-century revolution in calculus and Cauchy's role in bringing it about. The first historical problem will be to explain the new nineteenth-century recognition of the mathematical importance of the foundations of analysis. Chapter 2 will document the general lack of interest in rigor in the calculus on the part of eighteenth-century mathematicians and try to account for that lack of interest. It will examine the types of definitions these mathematicians gave for the concepts of the calculus and, finally, will trace the gradual reawakening of interest in foundations at the end of the eighteenth century.

In the heart of this book, chapters 3 to 6, I shall examine a number of specific achievements of eighteenth-century mathematics and show how Cauchy transformed them into the basis of his rigorous calculus. I shall be especially interested in the development of the algebra of inequalities; the history of the concept of limit; the work done on the notions of continuity and convergence; and some eighteenth-century treatments of the derivative and the integral. I shall conclude the book with an assessment of the magnitude of Cauchy's achievement when viewed in its full historical setting.

Though I shall be looking constantly for the antecedents of Cauchy's work, I shall insist throughout on the creativity and originality of his accomplishments. Like another major innovator, Copernicus, Cauchy owed much to his predecessors. But, also like Copernicus, Cauchy contributed a change in point of view so fundamental that his science was transformed when he was done.

1 Cauchy and the Nineteenth-Century Revolution in Calculus

The Nineteenth-Century Ideal

When a nineteenth-century mathematician spoke about rigor in analysis, or in any other subject, he had several general things in mind. First, every concept of the subject had to be explicitly defined in terms of concepts whose nature was held to be already known. (This criterion would be modified by a twentieth-century mathematician to allow undefined terms—that is, terms defined solely by the postulates they satisfy; this change is related to the late nineteenth-century tendency toward abstraction in mathematics, an important development but beyond the scope of this book.) Following Weierstrass, in analysis this meant that every statement about equality between limits was translatable, by well-defined rules, into an algebraic statement involving inequalities. Second, theorems had to be proved, with every step in the proof justified by a previously proved theorem, by a definition, or by an explicitly stated axiom.[1] This meant in particular that the *derivation* of a result by manipulating symbols was not a *proof* of the result; nor did drawing a diagram prove statements about continuous curves. Third, the definitions chosen, and the theorems proved, had to be sufficiently broad to support the entire structure of valid results belonging to the subject. The calculus was a well-developed subject, with a known body of results. To make the calculus rigorous, then, *all* previous valid results would have to be derived from the rigorous foundation.

Many nineteenth-century mathematicians believed themselves superior to their eighteenth-century counterparts because they would no longer accept intuition as part of a mathematical proof or allow the power of notation to substitute for the rigor of a proof.[2] To be sure, even nineteenth-century mathematicians often pursued fruitful methods without the maximal possible rigor, especially in developing new subjects, and individual mathematicians differed in the importance they gave to foundations. Cauchy himself was not consistently rigorous in his research papers. Nevertheless, criteria like those listed above were constantly in Cauchy's mind when he developed his *Cours d'analyse*. When Cauchy referred in that work to the rigor of geometry as the ideal to which he aspired, he had in mind, not diagrams, but logical structure: the way the works of Euclid and Archimedes were constructed.[3]

Cauchy explicitly distinguished between heuristics and justification. He separated the task of discovering results by means of "the generalness of algebra"—that is, discovering results by extrapolating from finite symbolic expressions to infinite ones, or from real to complex ones—from the quite different task of proving theorems. He described his own methodological ideal in these words:

> As for methods, I have sought to give them all the rigor which exists in geometry, so as never to refer to reasons drawn from the generalness of algebra. Reasons of this type, though often enough admitted, especially in passing from convergent series to divergent series, and from real quantities to imaginary expressions, can be considered only ... as inductions, sometimes appropriate to suggest truth, but as having little accord with the much-praised exactness of the mathematical sciences.... Most [algebraic] formulas hold true only under certain conditions, and for certain values of the quantities they contain. By determining these conditions and these values, and by fixing precisely the sense of all the notations I use, *I make all uncertainty disappear*.[4]

These are high standards. Let us turn to Cauchy's work and see how he met them.

A First Look at Cauchy's Calculus

It is a commonplace among mathematicians that Cauchy gave the first rigorous definitions of limit, convergence, continuity, and derivative, and that he used these definitions to give the first essentially rigorous treatment of the calculus and the first systematic treatment of convergence tests for infinite series. Many people have heard also that Cauchy's rigorous proofs introduced delta-epsilon methods into analysis.[5] When the mathematician opens Cauchy's major works, he expects these beliefs to be confirmed. But on first looking into the *Cours d'analyse* (1821), he may be somewhat shocked to find no deltas or epsilons anywhere near the definition of limit; moreover, the words in the definition sound more like appeals to intuition than to the algebra of inequalities. Then, when he opens the *Calcul infinitésimal* (1823) to find the definition of a derivative, he may be even more surprised to find that Cauchy defined the derivative as the ratio of the quotient of differences when the differences are infinitesimal. (He may be surprised also at the word *infinitésimal* in the title.) Returning to the *Cours d'analyse*, the mathematician is apt to be disappointed with the treatment of convergence of series, for although Cauchy used the Cauchy criterion, he did not even try to prove that it is a sufficient condition for convergence. The mathematician may well conclude that Cauchy's rigor has been highly overrated.[6]

But the discrepancy between what the mathematician expects and what Cauchy actually did is more apparent than real. In fact Cauchy's definitions and procedures are rigorous not only in the sense of "better than what came before" but in terms of nearly all that the mathematician expects. One major difficulty the modern reader finds in appreciating Cauchy comes from his old-fashioned terminology, the use of which—as will be seen—was deliberate. Another source of difficulty is the fact that the two books of 1821–1823 were originally lectures given to students who planned to apply the calculus.[7] Finally, the modern reader is likely to be unfamiliar with the more discursive style of mathematical exposition used in the early nineteeth century. Once the *Cours d'analyse* or the *Calcul infinitésimal* has been examined more closely, it will be seen that Cauchy's achievement is as impressive as expected.[8]

To illustrate what we have just said, let us look at the central concept in Cauchy's analysis, that of limit, on which his definitions of continuity, convergence, derivative, and integral all rest: "When the successively attributed values of one variable approach indefinitely a fixed value, finishing by differing from that fixed value by as little as desired, that fixed value is called the *limit* of all the others."[9] This definition seems at first to resemble the imprecise eighteenth-century definitions of limit more than it does the modern delta-epsilon definition. For instance, a classic eighteenth-century formulation is that given by d'Alembert and de la Chapelle in the *Encyclopédie*:

> One magnitude is said to be the *limit* of another magnitude, when the second can approach nearer to the first than a given magnitude, as small as that [given] magnitude may be supposed; nevertheless, *without* the magnitude which is approaching *ever being able to surpass* the magnitude which it approaches, so that the difference between a quantity and its limit is absolutely inassignable.... Properly speaking, the limit never coincides, or *never becomes equal*, to the quantity of which it is a limit, but the latter can always approach closer and closer, and can differ from it by as little as desired.[10] [Italics mine]

But the important difference between Cauchy's definition and those of his predecessors is that when Cauchy used his definition of limit in a proof, he often translated it into the language of inequalities. Sometimes, instead of so translating it, he left the job for the reader. But Cauchy knew exactly what the relevant inequalities were, and this was a significant new achievement. For example, he interpreted the statement "the limit, as x goes to infinity, of $f(x+1)-f(x)$ is some finite number k" as follows: "Designate by ε a number as small as desired. Since the increasing values of x will make the difference $f(x+1)-f(x)$ converge to the limit k, we can give to h a value sufficiently large so that, x being equal to or greater than h, the difference in question is included between $k-\varepsilon$ and $k+\varepsilon$."[11] This is hard to improve on. (We will find a delta to go with the epsilon when we describe Cauchy's theory of the derivative in chapter 5.)

Moreover, Cauchy's defining inequalities about limits were not ends in themselves; their purpose was to support a logical structure of results about the concepts of the calculus. For instance, he used the inequality we have just mentioned in a proof of the theorem that if $\lim_{x\to\infty} f(x+1)-f(x)=k$, then $\lim_{x\to\infty} f(x)/x=k$ also,[12] and then used an analogue of this theorem as the basis for a proof of the root test for convergence of series.[13] By contrast, definitions like the one in the *Encyclopédie* were not translated into inequalities and, more important, were almost never used to prove anything of substance.

In addition, the *Encyclopédie* definition has certain conceptual limitations which Cauchy's definition eliminated. For instance, the eighteenth-century term *magnitude* is less precise than Cauchy's term *variable*; rather than have one magnitude approach another, Cauchy clearly distinguished between the variable and the fixed value which is the limit of the variable. Moreover, the sense of the word *approach* is unclear in the *Encyclopédie* definition. Though it could have been understood in terms of inequalities, d'Alembert and de la Chapelle, unlike Cauchy, did not explicitly make the translation. Probably instead they were appealing to the idea of motion, as Newton had done in explaining his calculus.

More important, d'Alembert's and de la Chapelle's two restrictions on the variable's approach to the limit, italicized in my citation of their definition, are too strong for mathematical usefulness. If a magnitude never *surpasses* its limit, then a variable cannot oscillate around the limit. How then could we use this definition to define the limit of the partial sums of the series $1 - 1/2 + 1/3 - 1/4 + \cdots$, or to evaluate the limit, as x goes to zero, of $x^2 \sin 1/x$? And if a magnitude can never *equal* its limit, how can the derivative of the linear function $f(x) = ax + b$ be defined as the limit of the quotient of differences? Abandoning these restrictions was necessary to make the definition of limit sufficiently broad to support the definitions of the other basic concepts of the calculus.

Cauchy's definition of limit, of course, has a history. But my main point here has been to exhibit the contrast between the usual understanding of the limit concept in the eighteenth century and that brought about by Cauchy. This contrast exemplifies both the nature and quality of Cauchy's innovations. It will be seen later how Cauchy applied his concept of limit to establish a rigorous theory of convergence of series, to define continuity and prove the intermediate-value theorem for continuous functions, and to develop delta-epsilon proofs about derivatives and integrals—in short, to provide an algebraic foundation for the calculus. (For a sample of Cauchy's work, the reader may consult the selected texts translated in the appendix.)

Cauchy and Bolzano

Cauchy's achievements, though outstanding, were not unique. His contemporary Bernhard Bolzano made many similar discoveries. Though Bolzano's impact on the mathematics of his time appears to have been negligible,[14] his work was nonetheless excellent. Regardless of his relative lack of influence, it will be worthwhile to be aware of some of his achievements. For if a key to understanding Cauchy is an understanding of the influence his predecessors had on him, the importance of these predecessors will be even clearer if they can be shown to have influenced Bolzano in the same ways. In fact, the simultaneous discovery involved in Bolzano's achievement provides us with a controlled experiment, enabling us to see both what could be accomplished in the 1820s and what currents in eighteenth-century mathematical work made Bolzano's and Cauchy's accomplishments possible. For the present I will compare Bolzano's achievements with Cauchy's; a detailed analysis of their common historical background will be given in later chapters.

Bolzano, like Cauchy, wanted to introduce higher standards of rigor into analysis. Bolzano's idea of what makes a proof rigorous is expressed by the very titles of some of his works: one was the "purely analytic proof" of the intermediate-value theorem for continuous functions,[15] while another dealt with length, area, and volume "without consideration of the infinitely small . . . and without any suppositions not rigorously proved."[16] Everything was to be reduced to algebra, without appeals to infinitesimals, geometry, the ideas of space and time, or any other intuitive ideas.[17] For Bolzano as for Cauchy, the algebra of inequalities played an important role in proofs. But Bolzano emphasized, much more than Cauchy, that he was breaking with the past. For instance, in contrast to Cauchy's conservative terminology, Bolzano deliberately avoided the language of motion and the term *infinitesimal.*

In 1817, Bolzano gave a definition of continuous function even more elegant in its statement than Cauchy's 1821 definition was to be. Bolzano then used this definition to prove the intermediate-value theorem for continuous functions. The proof, which was differently conceived from Cauchy's, used what we today call the Bolzano-Weierstrass property of the real numbers.[18] In addition, Bolzano's 1817 paper makes use of what we now call the Cauchy criterion for convergence of a sequence,[19] which Cauchy himself—independently, as I hope to show—stated as a criterion for the convergence of series in 1821.

Bolzano's accomplishments do not end here. In later papers, in which he acknowledged having read Cauchy's books, Bolzano did additional important work. He had an elegant, inequality-based treatment of many properties of the derivative, of extrema, and of the Taylor series.[20] A striking and original discovery of Bolzano's about derivatives was his example of a continuous function which was nowhere differentiable, a decisive blow to the intuitive picture of the behavior of continuous functions.[21]

It has been suggested that the coincidence between Cauchy's 1821–1823 works and Bolzano's 1817 paper are too great to have come about by chance. Ivor Grattan-Guinness has argued that Cauchy's entire program—reducing the calculus to *analyse algébrique*, giving inequality treatments of the basic concepts, establishing rigor—cannot be explained in terms of Cauchy's previous work. He explains the coincidence by assuming that Cauchy had seen Bolzano's paper, and used the material in it without giving Bolzano credit. He states also that Cauchy borrowed Bolzano's definition of continuous function, the Cauchy criterion, and the proof of the intermediate-value theorem for continuous functions.[22]

I do not find this accusation convincing,* but for my immediate purpose, it would not make much difference if it were true; my intention is to call attention to the importance of the eighteenth-century predecessors of both Cauchy and Bolzano. It is only because so little is known about the basis on which Cauchy built that his work could appear to be without prior influences. It is because, lacking a historical setting, we view the *Cours d'analyse* as founded on one man's genius alone that it is possible to claim that the genius was Bolzano.

But common influences suffice to explain the specific similarities between the work of Cauchy and Bolzano. The familiarity of Cauchy and Bolzano with the work of their

*The principal reason for my conclusion is the wealth of common prior influences; the rest of this book will document those influences in detail.[23] Also, simultaneous discoveries abound in the history of mathematics. In addition, as H. Freudenthal and H. Sinaceur have shown,[24] there are real conceptual differences between the work of Cauchy and the work of Bolzano.

common predecessors can be documented. Once this is done, the similarities between the work of Cauchy and Bolzano will point out strongly how important these predecessors were; in the period 1815–1825, any genius of sufficient magnitude, seeking rigor in analysis, could have done these things. There were two such geniuses.

Unfinished Business

No mathematical subject is ever perfected overnight. Neither Cauchy nor Bolzano had solved by 1825 all the outstanding problems of analysis. There were two major lacunae in Cauchy's work at this time. First, he—and also Bolzano—did not yet appreciate the distinction between convergence and uniform convergence or that between continuity and uniform continuity. Also, though Cauchy implicitly assumed several forms of the completeness axiom for the real numbers, he did not fully understand the nature of completeness or the related topological properties of sets of real numbers or of points in space.

In addition to these mathematical "errors," Cauchy sometimes made errors through haste or inadvertence, which he could have corrected himself according to his own principles. For instance, he gave an erroneous proof of the convergence of alternating series,[25] which can easily be corrected using the Cauchy criterion. More generally, his willingness to leave the working out of the explicit inequalities in many theorems as an exercise to the reader left ambiguities throughout his works.

The confusion between pointwise and uniform properties led to Cauchy's famous "proof," in 1821, of the false theorem that an infinite series of continuous functions is continuous;[26] in 1816, Bolzano too seems to have believed that an infinite series of continuous functions was continuous.[27] In 1826, Abel, in his study of the continuity of the sum of a power series, published a counterexample to Cauchy's false theorem, but Abel did not identify the error in the proof.[28] The elucidation of the difference between convergence and uniform convergence by men like Stokes, Weierstrass, and Cauchy himself was still more than a decade away.

The verbal formulations of limits and continuity used by Cauchy and Bolzano obscured the distinction between "for any epsilon, there is a delta that works for all x"

and "for any epsilon and for all x, there is a delta." The only tools for handling such distinctions were words, and the usual formulation with the word "always" suggested "for all x" as well as "for any epsilon."[29] None of the eighteenth-century inequality arguments on which Bolzano and Cauchy drew elucidated this distinction, so that it was not immediately clear how much depended on it. In chapter 5, it will be seen how Lagrange failed to distinguish between convergence and uniform convergence in a proof about derivatives and how Cauchy followed Lagrange throughout.

Again like his predecessors, Cauchy did not have explicit formulations for the completeness of the real numbers. Among the forms of the completeness property he implicitly assumed are that a bounded monotone sequence converges to a limit and that the Cauchy criterion is a sufficient condition for the convergence of a series.[30] Though Cauchy understood that a real number could be obtained as the limit of rationals, he did not develop this insight into a definition of real numbers or a detailed description of the properties of the real numbers.[31] For Cauchy, results like the monotone-sequence property appeared relatively obvious, and defining the real numbers or even elucidating the consequences of the monotone-sequence property did not seem an urgent task. After the properties of continuity, convergence, and uniform convergence were understood considerably better, the job of defining the real numbers and describing their properties and the related properties of point sets was done by men like Weierstrass, Heine, Méray, Cantor, and Dedekind.[32]

Conclusion

In spite of the problems he did not solve, Cauchy set into motion a complete reformulation of the basis of analysis. The impact of his work can be illustrated by examining the career of Niels Henrik Abel. Abel's reaction to the *Cours d'analyse* was almost like a religious conversion. Abel had begun his career in the standard eighteenth-century way, reading Euler and Lagrange and solving problems in their style. When Abel read the *Cours d'analyse*, he was immediately convinced that his previous work lacked rigor; indeed, he was amazed that eighteenth-century mathematicians had been able to do so much without having that rigor.[33] Such a change, almost seeing with new eyes, is characteristic of one who has experienced a revolution in thought. When Abel left his native Norway to travel to the mathematical centers of Europe, Paris was one of the important stops on his list. (Unfortunately, Cauchy's coldness to beginners somewhat discouraged Abel once he got there.)[34] Abel's admiration for the *Cours d'analyse* led him to use its methods and spirit in his own rigorous investigations of infinite series.

Cauchy's work was read and acted upon also by Bolzano, Dirichlet, and Riemann; in part directly, in part through men like Abel, it influenced Weierstrass.[35] Thus it influenced the leading analysts of the mid-nineteenth century. Cauchy's books were translated into other languages, as well as being widely read in French,[36] and textbooks were written based on Cauchy's approach; the best known was Moigno's.[37] As further evidence of Cauchy's influence, we may note the widespread use of his methods of proof—identifiably Cauchy's because of the use of the delta-epsilon notation he introduced. Finally, we may cite the historical comments of nineteenth-century mathematicians, whose testimony is valuable because they knew the men who had learned from Cauchy.[38]

Bolzano's work also could have served as a starting point for the rigorization of analysis. Cauchy was the man who taught rigorous analysis to all of Europe, however, while Bolzano's works went almost unread until the 1860s (see note 14). This is not only because of the magnificent clarity of exposition in Cauchy's books; the reasons are partly social and institutional. The Ecole Polytechnique in Paris, where Cauchy delivered his lectures, was the first and foremost scientific school in Europe. Most of the leading French mathematicians and mathematical physicists of the age went there; many leading mathematicians read the courses of lectures written there as soon as they were published as books. Paris was the center of the mathematical world, and many mathematicians not lucky enough to be French came there to study.[39] In comparison, Bolzano worked in relative isolation in Prague, did not hold an important teaching position, published many of his papers as separate pamphlets because he had no ready access to a prestigious journal, and, finally, was known as a philosopher and theologian, rather than as a mathematician.[40]

Cauchy's enormous influence gives us another reason to want to know where his ideas came from. He synthesized previous work and built a firm foundation so well as to obscure the attempts that preceded him. Just as Euclid's *Elements* were so successful that they drove much earlier work into obscurity; just as the Newton-Leibniz calculus made it unnecessary to read much earlier work on areas and tangents; so Cauchy's *Cours d'analyse* and *Calcul infinitésimal* made obsolete many of the earlier treatments of limits, convergence, continuity, derivatives, and integrals.

After Cauchy, foundations had become an essential part of analysis, and Cauchy's books and teaching were largely responsible. Thus, explaining the transformation from the eighteenth-century calculus oriented toward results to nineteenth-century rigorous analysis means understanding the basis of Cauchy's work. I have argued that this transformation was revolutionary and have claimed that it requires an explanation. I shall now try to provide that explanation.

2 The Status of Foundations in Eighteenth-Century Calculus

Introduction

Mathematicians of the eighteenth century saw before them vast new worlds to explore and to conquer. In inventing the calculus, Newton and Leibniz had forged an algorithm of incredible power. Leibniz, in particular, chose his notations *d* and $\int$ for the basic, mutually inverse operations of his calculus precisely so that they could be applied in an almost automatic fashion.[1] Exploiting the new methods was the most exciting task in mathematics—and the most fruitful. One might think that an age so interested in the calculus would have given a high priority to setting the calculus on a rigorous basis. Yet a century and a half elapsed between the time the calculus was invented and the time Cauchy successfully gave it a logically acceptable form.

In attempting to understand why it took so long to make the calculus rigorous, one should not forget the difficulty of the task. To make a subject rigorous requires more than just choosing the appropriate definitions for the basic concepts; it is necessary also to be able to prove theorems about these concepts. Developing the methods needed for these proofs is seldom merely a trivial consequence of choosing the right definitions; in fact, the prior existence of the methods of proof is often necessary in order to recognize suitable definitions. A great deal of labor was required to devise the techniques and concepts needed to establish a firm foundation for the diverse results and applications of the calculus. It must also be noted, however, that progress in understanding the foundations of the calculus was relatively slow in the eighteenth century in large part because the interests of mathematicians lay elsewhere.

In the eighteenth century the problems considered to be most important were those which could be treated without paying attention to the foundations of the calculus. No strict line was drawn between the calculus and its applications, between mathematics and mathematical physics. Many of the results obtained in the calculus had immediate physical applications; this circumstance made attention to rigor less vital, since a test for the truth of the conclusions already existed—an empirical test. Moreover, the concentration on results made the fruitfulness of a well-chosen notation seem preferable to the more certain, but slower, theorem-and-proof procedure characteristic of Greek geometry. The century was dominated by a few enormously productive mathematicians—the Bernoullis, d'Alembert, Euler, Lagrange, and Laplace—whose work, with the exception of Lagrange, largely exemplified these tendencies. It was not until the end of the eighteenth century that the foundations of the calculus came to be recognized by leading mathematicians as a respectable mathematical problem.[2]

This chapter will describe and document the prevailing attitudes toward rigor in eighteenth-century analysis. In particular, it will ask and answer four questions. (1) What was the usual estimate in that period of the importance of the foundations of the calculus? It will be shown that the pursuit of rigorous foundations was essentially irrelevant to the major goals of the analysts of that time. Nevertheless, since rigor was sometimes discussed, we must ask, (2) What motivated that period's mathematicians when they did consider the foundations of the calculus? There was one major motivating factor from within mathematics: the desire to emulate the rigor of Greek geometry. But there were other motivations, some from outside mathematics. The history of eighteenth-century calculus cannot be understood in isolation from the general history of thought and of society.

In order to generalize meaningfully about the discussions of foundations that did occur, it is essential to have some idea of their content. Thus the following points must be considered: (3) When eighteenth-century mathematicians did treat the foundations of the calculus, what did they actually say? This question will be treated here only insofar as a general idea of the answer is needed[3] and also to indicate how the period's treatments of foundations fared at the hands of contemporary critics. It will be shown that although there were many explicit discussions of the basic concepts, these were seldom applied to nonelementary parts of analysis. This latter circumstance will be additional evidence that eighteenth-century analysts did not view making the calculus rigorous as their central concern. Finally, to find the background of Cauchy's rigorization, one must ask, (4) How and why did rigor come to be considered an important problem once again? It will be seen that Joseph-Louis Lagrange had the greatest role in restoring foundational questions to their central place in mathematics; Cauchy accepted Lagrange's estimate of the importance of foundations, though he gave a very different foundation. By understanding the prevailing eighteenth-century attitudes and how they changed only at the end of the century, it will be possible to understand what mathematicians of that time were actually doing when they created the techniques which Cauchy and Bolzano used to give the calculus a rigorous basis.

The Unimportance of Rigor in Eighteenth-Century Analysis

To determine the importance of rigor to eighteenth-century analysts, it will be necessary to look beyond the general statements made in the introductions to their books and examine their mathematics. To this end it will be useful to begin with a brief topical survey of the major types of research activity in eighteenth-century mathematics.[4]

Algebra was the theory of equations—especially the study of root-coefficient relations—and the study of methods of solving equations: solving them exactly, when possible; otherwise, by approximation processes, preferably ones that converged quickly. Only in 1770–1771 was the general question of solvability first attacked, by Lagrange. "The analysis of the infinite," a branch of mathematics named by Euler, involved finding the sums of infinite series and transforming them from one form into another, as well as finding the limits of infinite products and continued fractions; there were no attempts to formulate a general theory of convergence, though the speed of convergence of particular series was discussed and some convergence criteria were evolved. The differential calculus studied the finding of differentials of all orders, their mutual relations, and their applications to problems of geometry and physics. The integral calculus solved classes of differential equations and evaluated definite integrals, but did not prove the existence of solutions to either type of problem.

Differential equations were more often studied in the context of particular physical problems than in general mathematical theories.

Of course the relationships between these different areas of mathematics were noted, and methods developed for one type of problem often were applied to many others. But the single-minded search for general concepts and the establishment of rigorous foundations were not the major goals of eighteenth-century mathematics. These tendencies, it must be noted, were stronger on the Continent than in Great Britain, but it was on the Continent that the greatest amount of mathematics was done. Thus, it would be fair to describe much of the activity of mathematicians of this period as the exploitation of the power of symbols for the purpose of solving problems; their accomplishments here were outstanding.[5]

To get a sense of the content of these generalizations, we may sample the contents of a leading journal, the *Mémoires de l'Académie Royale des Sciences et Belles-Lettres de Berlin*, for the years 1750, 1775, and 1800, in the section *Classe de Mathématiques*. Under this heading for the year 1750, we find four articles by Leonhard Euler: one deals with deriving the equations of motion for a rotating body; another concerns the varying degree of light given off by the sun and other celestial bodies; the third is on the precession of the equinoxes; the fourth is on the effect of a hydraulic machine proposed by Johann Andreas Segner. There are two papers by Jean Le Rond d'Alembert, each the continuation of an earlier paper. The first treats the general form of the solution to the differential equation of the vibrating string. The second is about the integration of certain rational functions. In addition, there is a paper by Johann Kies on the brightness of Venus, and two papers determining the parallax of the moon and the curvature of the earth based on observations by J.-J. Lalande and a M. Chret. The topics of these papers show that the *Classe de Mathématiques* included not just mathematics, but mathematical physics and celestial mechanics. In the purely mathematical articles, the standards of proof are what might be termed formalistic, in the sense that algebraic derivations serve to establish the validity as well as the form of most of the results.[6] Worthy of note also are the domination of the journal by Euler and d'Alembert, and the fundamental nature of some of the questions treated in their papers: angular momentum, the vibrating string and the wave equation, the fundamental theorem of algebra. This volume is very much in the mainstream of eighteenth-century mathematics.

Returning to the same journal in 1775, there has been a change in personnel, but the situation is otherwise similar. There are three papers by Joseph-Louis Lagrange. The first deals with techniques for studying recurrent series and with integrating linear equations of finite partial differences, and applies these techniques to probability theory. The second treats the attraction of elliptical spheroids. The last paper is a contribution to number theory, asking when the quadratic form $py^2 + 2qyz + rz^2$ can be written in the form $4na + b$.[7] There are two papers by Johann (II) Bernoulli: one, devoted largely to trigonometric methods, discusses the position of the pole star; the other reports some observations of eclipses. There is, finally, one memoir by Nicolas Beguelin, investigating how many prime numbers there are greater or less than a given number. We note again the close relationship between mathematics and mathematical physics and the large number of significant articles in the journal by one major mathematician. The proofs in the mathematical articles are, again, usually either algebraic derivations or appeals to what is already known in analysis.

For the Berlin *Mémoires* of 1800 the picture is slightly different. Among seven mathematical articles published, there is one article on the foundations of the calculus. This is the second part of an essay by J.-P. Gruson, "Le calcul d'éxposition."[8] The other six papers, one by Abel Burja, one by Johann Elert Bode, and four by Jean Trembley, resemble in their subjects the material we have already considered: the length of a pendulum with given period at Berlin; astonomical observations; integrating equations involving finite differences; the attraction and equilibrium of spheroids; statistics about the duration of marriages; the precession of the equinoxes. Note in passing that the Berlin *Mémoires* is no longer the leading scientific journal it had been; with the death of Frederick the Great, who had patronized the Berlin Academy, the center of mathematics definitely shifted to Paris. Nevertheless, the topics covered in the Berlin *Mémoires* in its 1799–1800 volume are not atypical of the mathematics of that time. The one contribution to the foundations of the calculus does illustrate a slight shift in mathematical interest; Gruson's work is closely akin to that of Lagrange, a fact which is consistent with the influence I shall show for Lagrange later in this chapter.[9]

Examining other scientific journals of the eighteenth century for contributions to the foundations of the calculus provides analogous results.[10] Similar generalizations can be supported by reading through the collected papers of Euler, d'Alembert, Johann Bernoulli, or Laplace; though the standards of rigor used vary, by and large rigor is not a central concern. There is nothing reprehensible about the period's attitude toward foundations, even though it is not ours. For instance, treating infinite series as if they were polynomials had led to important new results without a general theory of convergence; why then should mathematicians not go on treating infinite series in that way? Such a procedure is not, at least in principle, inadmissible.[11] Similarly, since differential equations usually arose out of physical problems, there was no need to consider the problem of existence of solutions; the physical reality guaranteed that a solution existed. Complex numbers were obviously useful in algebra and trigonometry even though $\sqrt{-1}$ had only a formal definition; why not use the fruitful formulas containing $\sqrt{-1}$, since they led to valid conclusions?

Nor was this indifference to foundational questions a matter of hostility. Eighteenth-century mathematicians were willing to discuss basic questions when settling them was necessary to solve a problem. In trying to decide what sort of functions solved the differential equation for the vibrating string, vigorous opinions were expressed over which functions were admissible in analysis. This debate, which involved d'Alembert, Euler, Lagrange, and Daniel Bernoulli among others, had some influence on the Cauchy–Bolzano definition of continuous function, and probably also on Cauchy's concept of the integral.[12] Although this debate was not undertaken in order to provide a foundation for the calculus, it can be viewed as a partial exception to the generalizations above; but in their defense I should add that the participants themselves did not push the discussion very far beyond the particular circumstances that gave rise to it.[13]

But what of the strongest modern argument for rigorous foundations—that they help avoid mistakes? One might think that the desire to correct errors or to resolve contradictions would have led eighteenth-century mathematicians to discuss the foundations of the calculus. But this was not the case. There was no 'scandal' demanding immediate attention: there were no contradictions serious enough to halt the progress of mathematics. Even in the light of modern knowledge, these mathematicians made surprisingly few errors. In part this was because the infinite series they treated were usually power series with bounded coefficients, which behave very much as the analogy with polynomials would lead one to expect, even in the absence of a general theory of convergence. Also, the functions that these mathematicians studied often arose from physical models and thus were relatively well behaved. Experience must have quickly shown that certain types of arguments led one astray and therefore simply should not be used.[14] Finally, we must acknowledge the great genius of these men, particularly Euler, whose incredible ability to choose fruitful methods of derivation led them past many potential pitfalls. In the absence of obvious errors in mathematical work, eighteenth-century mathematicians apparently did not feel one of the traditional attractions of greater rigor: the need to separate the true from the false.

Too often, histories of the calculus inadvertently have given the impression that eighteenth-century mathematicians spent a great deal of their time and effort discussing the foundations of the calculus. Of course there was a debate about foundations in that period, and this obviously is important in understanding the origin of later foundations. But keep in mind where most eighteenth-century mathematicians' interests lay when evaluating their work on foundations. Solving problems was important, not proofs about the concepts used in solving them. The attitude of most of these mathematicians, implicit in their usual choice of problems and methods, may be summed up in a remark attributed to d'Alembert: "Go on, go on; the faith will come to you."[15]

It is in this context that we must understand the history of the foundations of the calculus. It is not the story of the successive steps that ultimately and naturally led to the establishment of Cauchy's rigorous formulation. Instead, it should be viewed as the record of a debate that was seldom of vital concern to its participants. In fact, we need to explain why an eighteenth-century mathematician like Lagrange would return so many times in his long mathematical career to a problem—finding the foundations of the calculus and deriving the major results from those foundations—in which nobody else seemed greatly interested. The mere existence of an unsolved problem does not ensure that people will even try to find a solution, let alone succeed in solving it.

The Occasion and Purpose for Discussions of Foundations

If eighteenth-century mathematicians were not very interested in foundations, when and why did they discuss the subject at all?[16] Explanations by leading mathematicians of the nature of the calculus were found principally in introductions to expositions of the calculus, especially in courses of lectures and in books based on such courses, in popular expositions of mathematics for the lay public, and in responses to attacks on the logical soundness of the calculus. They are not usually found in papers printed in scientific journals. Until the 1780s, there are almost no exceptions to this statement.[17] Indeed, these generalizations continue to hold well into the following century.

The need to begin an exposition of a subject by defining its basic terms is both psychological and logical, especially in mathematics, which traditionally has had a Euclidean form. But there were additional reasons for introducing expositions of calculus with explanations of the fundamental concepts. First, calculus was a new subject; thus, even nonelementary expositions like Newton's *Method of Fluxions* contained introductory material explaining the basic concepts of the calculus. Second, these expositions were intended as textbooks for a growing readership. There was an increasing interest in the eighteenth century in mathematics and the sciences. On the one hand, the potential audience among scientific professionals for such expositions was greater in size than ever before. The founding of scientific societies and journals, which dates from the mid-seventeenth century, made science—including mathematics—an organized, ongoing enterprise.[18] Textbooks, both elementary and advanced, were needed by the new and growing scientific community; in papers written as contributions to the community of working mathematicians, we do not generally find men like Laplace or Euler laying the foundations for the calculus. On the other hand, there was a large nonprofessional audience for scientific works. Public interest in science and mathematics had been greatly increased by the success of Newtonian physics in understanding the laws of the universe, prompting both mathematicians and philosophers to explain the calculus to laymen. For instance, d'Alembert's contributions to the *Encyclopédie*, undertaken by the French *philosophes* to systematize the knowledge of the age, included explanations of the calculus completely divorced from any expositions of major results. These are part of the tradition of eighteenth-century popularized science that ranged from Maclaurin's *Account of Sir Isaac Newton's Philosophical Discoveries* to Algarotti's *Newtonianism for the Ladies*.[19]

The Newton–Leibniz controversy also contributed to discussion of the foundations of the calculus.[20] Newton and his followers stressed certain points about his calculus in an attempt to show that it was different from—and superior in rigor to—that of Leibniz. In trying to defend Newton's calculus, British mathematicians emphasized the superior rigor of geometry over the unrigorous—and algebraic, not geometric—infinitesimals of Continental mathematicians.[21] This goal led the British to extensive discussions of foundations.

Explanations growing out of actual teaching experience became more common at the end of the eighteenth century, by which time most of the mathematicians were teachers. In the eighteenth century, many mathematicians and scientists had depended upon royal patronage or on personal wealth for their support. There were few university positions. But as the scientific community grew, more men of the middle class became scientists; these men needed support. In addition, there was a growing conviction that scientists could be useful to a nation, both for its expanding industries and its military capabilities.[22] In response to these changes, new schools, and scientific departments in old schools, were opened. These increased the number of jobs available for scientists. The most important example of a new school founded in response to these changes was the Ecole Polytechnique in Paris, founded by the Revolutionary government in 1795. The school was not established out of an altruistic desire to give jobs to would-be scientists, although its policy of "open enrollment," which recognized talent over class origins, fit nicely with the ideals of the Revolution. The founders of the Ecole Polytechnique recognized that science and mathematics were valuable to the state and proposed to use the school in its service to recruit and train scientists and engineers. Other nations followed this example. Teaching, perhaps even more than writing textbooks, stimulated mathematicians to consider the foundations of their subject. In presenting a subject like analysis to beginners, no appeal could be made to the way a concept is understood in use, since the beginner did not have the experience needed for that understanding. Having students tends to force a teacher to expound the first principles of a subject clearly and to think those principles through anew. This helps to explain why the contributions to the foundations of the calculus of Lagrange, Cauchy, Weierstrass, and Dedekind were all stimulated by their teaching.[23]

The foundations of the calculus thus seem to have been viewed as a matter more philosophical or pedagogical than mathematical. Indeed, the phrase "true metaphysics of the calculus," not "basic axioms and definitions" as might be found in geometry, recurs in the titles of eighteenth-century discussions of foundations.[24] Yet there was also a desire among some mathematicians near the end of the century to call attention to foundations. Lagrange, for example, believed that his foundations were establishing the basis of a completed structure, since the calculus had fairly well succeeded in solving the major problems set for it. In fact, he once termed higher mathematics "decadent."[25] This attitude, which has been called "'fin du siècle' pessimism,"[26] has, when viewed from a modern perspective, some basis in fact. Without careful attention to convergence of series, without knowing under what conditions one may change the order of taking limits or integration, certain classes of results cannot be obtained. Denis Diderot, who was hostile to the mathematical method in science, nevertheless made an illuminating observation about the mathematics of his contemporaries—shared by at least some of them—when he said that men like the Bernoullis, Euler, and d'Alembert had "erected the pillars of Hercules" beyond which later ages would not be able to pass.[27]

A last reason for discussing the basic concepts of the calculus was that the rigor of the calculus had been attacked and mathematicians wanted to defend it. Though some early attacks had come from mathematicians who objected to infinitesimals, it is surely a measure of the lack of interest in foundations on the part of mathematicians that until the 1780s, the most prominent attacks on what was in fact logically inadequate came from philosophers and theologians.[28] The most telling and influential criticism of the calculus came in 1734 from George Berkeley, Bishop of Cloyne, and was undertaken partly to defend religion against attacks by scientists.[29] The cause for Berkeley's attack was not some personal pique against mathematicians; it was part of his opposition to the prevailing views of the Enlightenment.

Eighteenth-century philosophy drew on the prestige of Newtonian science and on the materialistic philosophies of Francis Bacon, Robert Boyle, Newton, and John Locke. Newton had argued that the perfect order of what was later called "the Newtonian world-machine" proved that there was a God who had created it. But it was a long step from Newton's lawful nature and the corresponding "Nature's God" to the God of orthodox Christianity. So along with the widespread public interest in scientific matters and public respect for the achievements of mathematics and science, there was a growing iconoclasm directed against old philosophies, feudal governments, and the Christian religion.[30] Bishop Berkeley wanted to defend Christianity against attacks by scientists who, he said, falsely believed that their superior ability to reason made them better judges of religious matters than were Christian clergymen. Berkeley counterattacked, pointing out weaknesses in what Enlightenment thinkers thought was the most secure achievement of Reason: mathematics. His attack, *The Analyst, or a Discourse Addressed to an Infidel Mathematician*, ended with sixty-seven rhetorical queries, of which three will serve to summarize his motives—and his style:

Query 56:
Whether the corpuscularian, experimental, and mathematical philosophy, so much cultivated in the last age, hath not too much engrossed men's attention: some part whereof it might have usefully employed? . . .

Query 62:
Whether mysteries may not with better right be allowed of in Divine faith than in human sciences?

Query 63:
Whether such mathematicians as cry out against mysteries have ever examined their own principles? [31]

Berkeley's attack on the calculus pointed out real deficiencies, as we shall see. As the above quotations illustrate, his attack was also incisive, witty, and infuriating. Many mathematicians were moved to try to answer it. In fact, several important eighteenth-century discussions of the foundations of the calculus can be traced back to Berkeley's attack. For instance, Maclaurin's monumental two-volume *A Treatise of Fluxions* began as a reply to Berkeley. Berkeley's attack had a more lasting effect than simply stimulating an immediate set of replies; it served to keep the question of foundations alive and under discussion, and it pointed to the questions which had to be answered if a successful foundation were to be given. D'Alembert and Lazare Carnot both used some of Berkeley's arguments in their own discussions of foundations, and Lagrange took Berkeley's criticisms with the utmost seriousness.[32]

Mathematics often appears to be an autonomous, self-directing activity. It has an inner life and logic of its own to a greater degree than any of the natural sciences. Nevertheless, the outside world impinges upon it in many ways, not in determining answers but in influencing which questions are asked. And this tendency was especially marked in the history of the foundations of the calculus in the eighteenth century. External causes were not the only ones producing discussions of foundations; but had public interest in mathematics and science been less, and had mathematicians not been obliged to teach, there would have been even fewer such discussions. Though this situation is surprising in the light of modern views of mathematics, it is understandable in terms of the major interests of mathematicians in the eighteenth century. Inasmuch as no major errors had been found, there was little reason for them to act any differently. The traditional view of mathematics—self-evident assumptions, clear definitions, logically sound proofs—may apply to the geometry of the Greeks, but it does not describe eighteenth-century analysis.

Keeping Rigor Alive: The Persistence of the Greek Tradition

Algebra, not geometry, was the model for eighteenth-century mathematical practice.[33] For research mathematicians, '*analysis*'—problem solving—was prized over '*synthesis*'—proof.[34] But tnough the subject matter of Euclid's geometry was only a secondary area of mathematical concern, the logical structure of the works of Euclid and Archimedes was universally admired. Their example was the strongest force from within mathematics directing attention toward rigor. The persistence of the Greek tradition kept the old standards alive and in everybody's consciousness.

In this respect, mathematics, like the other sciences, was a part of the scientific revolution; a Greek tradition was present in all the new sciences of Renaissance and seventeenth-century Europe.[35] However, in no subject was the respect for the Greek tradition as great as in mathematics, and nowhere else was it as long lasting. Though Copernicus used Greek astronomical methods in his work, he made an earth-shaking change. Though Galileo said that Aristotle, had he been alive in the seventeenth century, would have immediately adopted the principles of Galileo's physics, a major part of Galileo's work is devoted to refuting the writings of Aristotle and his followers. But there is no figure in the mathematical part of the scientific revolution who played the role of Copernicus to Ptolemy or Galileo to Aristotle: nobody overthrew the old system in mathematics.[36] To be sure, algebra, analytic geometry, and calculus added immensely to mathematical methods and knowledge. Still, mathematicians had little fault to find with the proofs and conclusions of Greek geometry. The usual criticism of the Greeks was that they had not gone far enough; nobody attacked their achievements. (Not even the nineteenth-century originators of non-Euclidean geometry, Bolyai and Lobachevsky, overthrew Euclid; it was Voltaire, not Euclid, who claimed that there was only one geometry.) Respect for the lasting accomplishment of the Greeks kept the standards of rigorous mathematics more alive in the eighteenth century than the nature of mathematical practice required.

Thus one contribution of the Greeks to establishing rigor in the calculus was in having provided a model of rigorous reasoning, a model that influenced philosophers as well as mathematicians in their expectations about mathematics. But in addition, specific Greek theorems were important parts of eighteenth-century mathematics. Euclid's *Elements* contained a theory of irrational ratios[37] that was viewed as the ultimate foundation of arithmetic, and whose basic assumptions include what is now called the Archimedean axiom: "Quantities have a ratio when they are capable, on being multiplied, of exceeding one another"—in modern terms, for any a, b in an Archimedean ordered field, there is an integer n such that $na > b$.[38] Eudoxus, the originator of Euclid's theory of irrationals, and Archimedes had used a type of argument now known as '*the method of exhaustion*' to prove theorems about the values of curvilinear areas and volumes. The method of proof, perhaps more correctly designated by Dijksterhuis as "indirect passage to the limit,"[39] proves that some curvilinear area A has some value K by indirect proof. For instance, consider the proof by Archimedes that the area of a circle is equal to half the circumference times the radius.[40] Let the area of the circle be A; half the circumference times the radius, K. First assume $A > K$. Let $A - K = D$. Then, by inscribing regular polygons of four, eight, . . . , 2^n sides in the circle, the geometry of the situation and the Archimedean axiom allow one to show that the area of some inscribed 2^n-sided polygon P differs from the area of the circle A by less than any given quantity[41]—in particular, by less than D. So $P > K$. But the area of the polygon is equal to half the perimeter times the altitude to its side; the perimeter is less than the circumference of the circle, the altitude less than the radius, so $P < K$. This is a contradiction; therefore the area of the circle A cannot exceed K. Archimedes then assumed $A < K$ and showed that this, too, leads to a contradiction. Hence we must conclude that $A = K$, QED. *If* the area A exists, this method of proof is valid.

These Greek theories greatly influenced eighteenth-century discussions of foundations. Infinitesimals were attacked because they did not obey the Archimedean axiom;[42] occasionally, proofs of the correctness of specific results of the calculus were given by the method of exhaustion as a way of satisfying absolutely everyone about the correctness of the conclusions obtained by analytical means.[43]

The method of exhaustion, however, though rigorous, did not solve the problem of giving a rigorous basis for the calculus. The method of exhaustion is extremely difficult to apply in complicated cases, and proving each result of the calculus by its means would be a superhuman task. Also, the method was out of tune with the prevailing algebraic spirit of eighteenth-century calculus, because it is essentially geometric. Moreover, the method of exhaustion is not always applicable to calculating the values of areas. To be sure, the Greeks had methods for computing areas, but these methods were not rigorously based, and most of them in any case were not well known in the seventeenth and eighteenth centuries.[44] The power of the calculus to discover such results was so great and its success in solving complicated problems so striking that the Greek methods of discovery, even if they had been known, might well have been viewed as historical curiosities.

Cauchy himself said that the Greek methods were his model of rigor, and we must take this statement seriously. Cauchy, like Euclid and Archimedes, gave his work a logical structure, basing chains of theorems on his definitions. In addition, Cauchy used techniques that have some kinship with Greek techniques. But Cauchy's techniques owed more to eighteenth-century algebra than to the method of exhaustion. The method of exhaustion was indeed one source of the simple limit arguments used in eighteenth-century analysis. These limit arguments—as in the work of L'Huilier, for instance—were often nothing but algebraic translations of Greek arguments, a fact which helped give the inequality-based limit arguments some degree of respectablity. But much more sophisticated inequality techniques and many results in analysis unknown to the Greeks were needed before the calculus could be founded on the algebra of inequalities. Furthermore, the methods of proof would have to be applied to all examples, not just elementary ones; only then would the techniques be a part of mathematics proper, rather than part of the explanation designed for beginners or critics.

The Debate over Foundations

In order to understand the climate of opinion before Lagrange was able to propose, and Cauchy to solve, the problem of providing adequate foundations for the calculus, there must be an understanding of eighteenth-century attempts to explain its concepts.[45] These explanations had deficiencies. They will not be criticized from the modern viewpoint; criticisms raised by contemporaries against these explanations, however, will be reviewed.

The calculus in the eighteenth century had two basic concepts: differential quotient and integral. The integral was not only known to be the inverse of the differential quotient; it was usually also *defined* as that inverse. Thus all the applications of the calculus—areas, volumes, solution of differential equations, tangents, maxima and minima, variational problems—rested on translating the given problem into an analytical or algebraic expression that could either be differentiated or shown to be the differential quotient of some other expression. Therefore any foundation for the calculus had to begin by justifying the rules for differentiating algebraic quantities.[46] The types of explanation may be surveyed by treating a very simple case: calculating the differential quotient of the function $y = x^2$. Taking such a simple case does not do violence to the spirit of eighteenth-century calculus; since its discussions of foundations were usually meant to illustrate the nature of the concepts of the calculus, not to justify them in full technical detail, it too usually considered simple cases.

All eighteenth-century mathematicians would begin this calculation in essentially the same way. If $y = x^2$ and if

x takes the value $x + h$ for some small h, y becomes $x^2 + 2xh + h^2$. If the change in x is h, the change in y is $2xh + h^2$ and the ratio of the changes is $(2xh + h^2)/h = 2x + h$. The differential quotient of x^2 (or in equivalent terminology, the fluxion or derivative of x^2) is not $2x + h$, however, but $2x$. What happened to the h? To be sure, h can be taken as small as desired. Still, how can completely neglecting it in the final expression be justified? Even the relatively uninterested analyst could not avoid fundamental questions of this sort. The problem was solved for differential quotients of arbitrary functions only by Cauchy, though it had been discussed throughout the eighteenth century. Here are the principal answers given then.

Infinitesimals Considering h to be an infinitesimally small quantity, we can neglect it with respect to the finite quantity $2x$. Thus $2x + h$ and $2x$ are essentially equal—or simply, are equal. This procedure was used by Leibniz in his early papers, and pursued by the Marquis de l'Hôpital and Johann Bernoulli. But this explanation was abandoned soon, because two major objections were raised against it. First, as Newton pointed out, infinitesimals did not obey the Archimedean axiom and therefore had no legitimate mathematical status. Second, unless $h = 0$, discarding it is an error, and as both Newton and Berkeley observed, "errours, tho' never so small, are not to be neglected in Mathematicks."[47]

Fluxions Newton and his British followers, of whom the most eminent was Maclaurin, took the velocities, or rates of change, of x and y to be their basic concepts. These velocities were finite quantities, and therefore apparently not subject to the objections raised against infinitesimals. Quantities like x and y, which are subject to change, were called *flowing quantities* or *fluents*; their velocities, written as $\dot{x}$ and $\dot{y}$, were called *fluxions*. For sufficiently small time intervals, the increments of x and y were assumed proportional to the velocities or fluxions with which they change; the increments in small time intervals could thus be defined in terms of those velocities. Thus if the change in x is $\dot{x}h$ for some "indefinitely small" h, the change in y is $\dot{y}h$. The ratio of the increments, $\dot{y}/\dot{x}$, is then computed to be $2x + h$. But since (in Newton's words) h is conceived to be "indefinitely little," it will be "nothing in respect of the rest"; it will vanish, and the ratio $\dot{y}/\dot{x}$ will be equal to $2x$.[48] But should the basic concepts of the calculus really be understood in terms of velocity? Bishop Berkeley, and after him d'Alembert and Lagrange, all pointed out that we had no independent idea of velocity clear enough to serve as a foundation for the calculus. Besides, velocity is an idea from physics; d'Alembert and Lagrange both insisted that the calculus should not depend on an idea outside mathematics.[49] As for the "indefinitely little," Newton himself later preferred to speak of limits or, as he also called them, "last" or "ultimate" ratios.

Early Limit Concept Since we are considering the ratio $2xh + h^2/h$ as h gets smaller and smaller, we may define $2x$ to be the limit of the ratio $2xh + h^2/h$ as h goes to zero; a *limit* is defined as a quantity which the ratio can never surpass, but can approach to within any given difference. This type of explanation was given by Newton. Newton spoke of the limit in another way also: as a "last ratio" or an "ultimate ratio." That is, the limit or ultimate ratio was the value of the ratio at the last instant of time before h—an "evanescent quantity"—has vanished.[50] In this instant, the ratio became equal to $2x$. Similarly, at the last instant, the secant coincided with the tangent. The limit concept, as defined above, was also used in explaining the calculus by d'Alembert and by Silvestre-François Lacroix, and by Colin Maclaurin in discussing the sum of an infinite series.[51]

The criticisms raised against this version of the limit concept by Bishop Berkeley were almost impossible to answer in eighteenth-century terms. Berkeley's principal objection to the methods of limits and of ultimate ratios was that $2x + h$ can *never* actually become equal to $2x$. Similarly, a secant can never *become* a tangent, no matter how close together the points of intersection are. Furthermore, the whole procedure used to find limits was illegitimate. Either h is equal to zero or it is not. If $h = 0$, then it is impossible to compute the ratio $2xh + h^2/h$, since if $h = 0$, then $(x + h)^2 = x^2$, not $x^2 + 2xh + h^2$. Also, if $h = 0$, we cannot divide both terms of the ratio $2xh + h^2/h$ by it to obtain the form $2x + h$. Nor can we evade these difficulties by saying that h is not equal to zero. If h is not zero, then $2x + h$ is never equal to $2x$. There is apparently no way out of this dilemma: if $2x + h$ does not actually equal $2x$, our result is inexact; if $2x + h$ does become equal to $2x$, our original ratio was $0/0$ and the whole argument is absurd. Berkeley emphasized that the quantity h "cannot be got rid of." With devastating effect, he borrowed Newton's stricture against infinitesimals to attack the method of limits: "errours, tho' never so small, are not to be neglected in Mathematicks."[52] Berkeley's strictures against these arguments were elaborated upon by Lagrange, who made even more convincing the proposition that the limit concept as it had existed in the eighteenth century could not provide a foundation for the calculus.

Compensation of Errors Although $2x + h$ is not equal to $2x$, the calculus nevertheless got correct results. For instance, the slope of the tangent to the parabola $y = x^2$ is indeed $2x$. But this correct result is obtained, according to Bishop Berkeley, because in applying the rule for finding the differential quotient of x^2 to the geometrical problem of tangents, an error exactly canceling the neglect of h was made: the slope of the tangent was determined by treating the curve at the given point as though it coincided with the tangent at two separate points, which it does not. Berkeley proved that these "errors" exactly canceled each other in the case of the tangent to $y = px^2$,[53] using facts known about the parabola and its tangent from Apollonius's *Conics* and relying only on finite arguments. Lazare Carnot later claimed that he could show that the errors made in procedures like these always canceled each other out.[54] But Berkeley's purpose had been to show that the calculus had *no* unexceptionable foundation; two errors are, after all, worse than one. He might have been amused to see Carnot try to convert that criticism into a valid basis for the calculus. Unfortunately, as Lagrange pointed out, it did not seem possible to prove that the errors made always canceled each other out.[55]

Greek Style The calculus can be viewed solely as a method of discovery whose results may be proved by other methods—by methods equivalent to the method of exhaustion of Greek geometry. In our example, let the value

$2x$ be given. Suppose now that the rate of change of y, which will be denoted R, is *not* equal to $2x$. Suppose first that $R > 2x$. Then, by assuming that for small h the rate of change is proportional to the increment in x, we can choose h sufficiently small to show that $R > 2x$ leads to a contradiction. Again assuming that for h small enough, the rate of change is proportional to the increment in x, it can be shown that $R < 2x$ leads to a contradiction. Hence $R = 2x$. This type of argument, modeled on Greek proofs about the equality of areas, was used extensively by Colin Maclaurin.[56] Maclaurin believed that one could go even further and prove the validity of all the applications of the calculus by appeals to the method of exhaustion. Given an acceptable definition of R, the rigor of Maclaurin's method would be beyond question. However, the rate of change R must somehow be defined, and the only concepts at Maclaurin's disposal were those of limit and fluxion, concepts whose drawbacks have already been discussed. Also, the technical demands of the proofs by the method of exhaustion are formidable in complicated cases.

Zeros The quantity h can be made less than any given quantity; hence, when added to a finite quantity like $2x$, it is actually zero. But although h is zero when *added* to finite quantities, it is not zero when considered in *ratios*; thus $(2xh + h^2)/h$ is not the meaningless $0/0$, but the well-defined $2x$. This explanation was given by Leonhard Euler and is similar to that later used by Laplace. Lagrange said that this method is equivalent to that of limits,[57] and therefore both his (and Berkeley's) criticisms apply to it.

Algebraic Method The equation $(x + h)^2 = x^2 + 2xh + h^2$ is a special case of the general equation

$$y(x + h) = y(x) + hp(x) + h^2q(x) + h^3r(x) + \cdots$$

for the case $y = x^2$. We *define* the coefficient $p(x)$ of h in the general expansion to be the derivative of y with respect to x. Since for our example $p(x) = 2x$, $2x$ is the derivative. This method was used by John Landen, L. F. A. Arbogast, J.-P. Gruson, and most influentially, by Lagrange.[58] Its advantage is that the derivative is *exactly* $2x$; we do not need to explain what happened to h. What must be explained, however, is how this definition corresponds to the usual one, especially in applications to finding tangents and rates of change, and how it is known that there is such a series expansion when the function y is not a rational function. Another problem is the uniqueness of the series expansion corresponding to the function; Cauchy effectively destroyed Lagrange's method by finding a class of counterexamples to its basic, if implicit, assumption that no two functions have the same Taylor series; Cauchy pointed out that, for example, the two distinct functions e^{-x^2} and $e^{-x^2} + e^{-1/x^2}$ have the same Taylor series about the point $x = 0$.[59]

Limits: Improved Version In the nineteenth century, Cauchy established a new meaning for the statement "$2x$ is the limit, as h goes to zero, of $2x + h$": it means *exactly* that we can make $2x + h$ and $2x$ differ, in absolute value, by less than any given quantity if we take h sufficiently small. It does not matter whether the variable actually reaches or surpasses its limit. (Cauchy's limit concept has predecessors—for instance, Simon L'Huilier's explicit acceptance that the partial sums of a convergent alternating series could "surpass" their limit; see chapter 4.)

These, then, were the principal ways of explaining the concepts of the calculus from its invention in the seventeenth century to its rigorization by Cauchy in 1823. But there were more important differences between the eighteenth and nineteenth-century mathematicians than their explanations of the derivative of $y = x^2$. The former did not strive to apply such explanations to many nonelementary examples. For instance, the restriction on "never surpassing the limit" continued throughout the eighteenth century, even though many counterexamples—as well known as the series for log 2 or Leibniz's series for $\pi/4$ were known. Most mathematicians of the period seem to have believed that the details in nonelementary contexts could be supplied if the need arose; but they did not expect the need to arise often. They were in fact able to avoid major errors, though they could not treat some problems that became important later. Finally, these mathematicians did not base new results on foundations, nor did they expect to. Their foundations were not really able to systematize even the existing level of knowledge. Mathematical progress in the century was great, but it was made in areas other than the foundations of analysis.

The Revival of Interest in Foundations: Lagrange

The first major mathematician to treat the foundations of the calculus as a serious mathematical problem was Joseph-Louis Lagrange. Not only was he concerned with the problem of foundations, but unlike his contemporaries, he returned to it again and again. This repeated inquiry came in part because of the requirements of his teaching, in part because he became so convinced of the validity of Berkeley's criticisms that he could not remain content with the existing foundations. He was impressed, as were many others, by the Euclidean tradition. In addition, he thought that the body of results of the calculus was more or less complete and in need of systematization. In 1797, Lagrange published a book, the *Fonctions analytiques*, which claimed to have solved the problem of putting the calculus on a rigorous basis. Lagrange's book differed markedly from earlier expositions of the calculus. To be sure, his work, like those of others, gave definitions of the basic concepts. But, what was infinitely more important, he derived the existing major results from his foundation. The contribution of the *Fonctions analytiques* to the rigorization of analysis lies not in Lagrange's specific definition of the derivative as the coefficient of the linear term in the Taylor series, but in other things: in taking rigor seriously; in demolishing most older views; in the example he set of thoroughly working out the received results of the calculus from his definitions; and in developing the techniques necessary to carry this out.[60]

Though Lagrange's interest in the question of foundations was unusually long lasting,[61] he began his work on it for the usual eighteenth-century reasons. His first mention of the foundations of the calculus arose out of the need to teach. In a letter to Euler in 1754, Lagrange stated that he had worked out the elements of the differential and integral calculus for the use of his pupils at the military school in Turin.[62] He even claimed "to have developed the true metaphysic of their principles, in so far as this is possible." But at this time Lagrange shared the general view of the lack of importance of foundations and did not even bother to relate the details to Euler.

In 1760, Lagrange took up the subject again, this time to clarify a philosopher's description of the calculus. The philosopher was the Barnabite friar Hyacinth Sigismund Gerdil, and his view, held in opposition to that of Fontenelle, was that there was no actual infinite in mathematics—a conclusion which Gerdil believed to have important implications for philosophy.[63] One place where mathematicians apparently needed the actual infinite was L'Hôpital's determination of the asymptotes to a hyperbola by considering the asymptotes as tangents at infinity. Gerdil denied that the infinite was really involved in this argument, even though L'Hôpital had obtained correct results by assuming that it was.[64] Lagrange thought a further explanation necessary and gave it in a footnote to Gerdil's paper; for, while Gerdil principally had wished to show that the calculus did not need infinity, Lagrange (who agreed) wanted also to give his own reasons for believing that the calculus was nevertheless rigorous. Lagrange said that L'Hôpital had obtained the right answer only because the error made in assuming the hyperbola and its asymptotes met was compensated for by an equal and opposite error in treating differentials as though they were zero.[65] This circumstance in fact applied to all other uses of infinitesimals in the calculus; true results, said Lagrange, were obtained only by a mutual cancellation of errors, but nobody could *prove* that the errors always canceled out. Nevertheless, Lagrange added, the calculus could still be justified: the Newtonian calculus was entirely rigorous. To illustrate this, Lagrange briefly sketched the principles of the method of first and last ratios.[66]

Lagrange Takes a New View: The Calculus as Algebra

Between 1760 and 1772, Lagrange changed his views about the sufficiency of the Newton–d'Alembert foundation for the calculus. In a paper published in 1772, he declared instead that the concepts of the calculus could be made rigorous only if they were defined in terms of algebraic concepts—that is, the calculus had to be reduced to algebra.[67] This view of Lagrange's ultimately prevailed with Bolzano and Cauchy. The important question for now is, Why did Lagrange reject the earlier views? He did not give explicit reasons until much later, but he may have become dissatisfied with the standard foundations—

infinitesimals, fluxions, and limits—through Berkeley's criticisms.[68] He clearly did not believe that the older foundations measured up to the standards of reasoning expected in mathematics. He believed that a suitable standard was met with only in algebra.

On reading Lagrange's work, one is struck by his feeling for the general: his idea that any particular result of any interest, whether it is the unsolvability of the quintic equation, the accuracy of an approximation, or the equations of motion for a physical system, is—and must be shown to be—a special case of some more general principle.[69] His extreme love of generality was unusual for his time and contrasts with the emphasis of many of his contemporaries on solving specific problems. His algebraic foundation for the calculus was consistent with his generalizing tendency.

Lagrange first explained his new idea for the algebraic foundations of the calculus in a paper of 1772, "Sur une nouvelle espèce du calcul rélatif à la différentiation et à l'intégration des quantités variables."[70] He believed that there was a wholly algebraic "theory of series" which gave any function a power-series expansion;[71] it was on this idea that his new foundation rested. "The differential calculus, considered in all its generality," he said, "consists in finding directly, and by easy and simple procedures," the functions $p, p', p'', \ldots$ in the general expansion

$$u(x+h) = u(x) + ph + p'h^2 + p''h^3 + \cdots$$

for a given function $u(x)$.* Lagrange said that this view of the calculus was "the clearest and simplest yet given." He was thus consciously and explicitly breaking with the earlier tradition. His new foundation was intended to be purely algebraic, untainted either by philosophy or by

* Lagrange termed p the "derived function" of $u(x)$; this term is the origin of our derivative. The second derivative of $u(x)$ he defined as the derivative of p, that is, as the coefficient of h in the expansion of $p(x+h)$, and so on recursively. He also introduced the familiar notation $u'(x), u''(x), \ldots$ for the successive derived functions of $u(x)$, writing the Taylor series $u(x+h) = u(x) + hu'(x) + (h^2/1 \cdot 2)u''(x) + \cdots$.

fuzzy ideas, "independent of all metaphysics and all theory of infinitely small and evanescent quantities."[72]

Lagrange thereby became the first major mathematician to grant the validity of the criticisms of the old foundations of the calculus. Nevertheless, in 1772 Lagrange still had not broken completely with the prevailing attitude toward foundations; it was not until 1797 that he gave in print explicit reasons for rejecting the earlier views. Foundations were not the primary subject of his 1772 paper, and his remarks concerning them were brief and incidental. He did not at this time even deduce the basic algorithmic rules of the differential calculus from his new foundation, let alone give such deductions for more complicated results.

Lagrange did think it important to find an exact and algebraic foundation for the calculus, however, and he continued to think this for the rest of his life. But in 1772 it was still an embryonic idea, not a realized goal. He was not yet satisfied with his own foundation; we know this because, through the medium of the Berlin Academy of Sciences' prize competition in 1784, he and his colleagues appealed to the entire learned world to solve the problem of the foundations of the calculus.

Stating the Problem: The Berlin Prize Competition

In the eighteenth century, learned academies often offered prizes for the solution of outstanding scientific problems. One purpose of these prize competitions was to attract the attention of mathematicians to a major question and get it solved. Solving such a problem could make a mathematician's reputation.[73] We can get some idea of what the prize questions were like by returning to the Berlin Academy's journal for the years 1750 and 1775. In 1750, the question concerned the resistance of fluids to motion; in 1774–1775, perfecting the methods used in computing the orbits of comets. In 1784, at the suggestion of Lagrange, the Berlin Academy proposed the question of the foundations of the calculus as the mathematical prize problem.[74] The date 1784 can be taken to mark Lagrange's public recognition of two propositions: first, the foundations of the calculus were unsatisfactory; and second, this situation posed a major unsolved mathematical problem. The traditional attitudes would no longer do; Lagrange wanted mathematicians to consider this problem, with which he had been concerned for so long, with greater seriousness. Though it is hard to document the effect of Lagrange's public admission, it certainly had some effect.

Two full-scale books by mathematicians devoted to expounding the foundations of the calculus and deriving the existing structure of results from these foundations were based on essays submitted in this contest: L'Huilier's *Exposition élémentaire* (1787) and Lazare Carnot's *Réflexions sur la métaphysique du calcul infinitésimal* (1797); since these and Lagrange's own *Fonctions analytiques* (1797) were the only such books published on the Continent in the eighteenth century, Lagrange was directly responsible for all the major manifestations of the new interest in foundations as a respectable mathematical problem.

The tone of the Berlin prize proposal was concern, not complacency. The academicians recognized that they were asking for something that did not yet exist. It was not just a matter of finding plausible definitions of the basic concepts but of explaining the success of the calculus as it then existed—the wealth of received results and powerful methods. Nineteenth-century mathematicians often wondered how the vast structure of analysis could have been erected on the inadequate foundations of the previous century. It was a good question, and the Berlin prize proposal marks the first real recognition of it.

The prize was to be awarded for "a clear and precise theory of what is called *Infinity* in mathematics." The proposal began by saying that such a theory was needed so that mathematics could continue to be respected for its rigor and its precision. The proposal then became more specific:

It is well known that higher mathematics continually uses infinitely large and infinitely small quantities. Nevertheless, geometers, and even the ancient analysts, have carefully avoided everything which approaches the infinite; and some great modern analysts hold that the terms of the expression "infinite magnitude" contradict one another.[75]

The Academy hopes, therefore, that it can be explained how so many true theorems have been deduced from a contradictory supposition, and that a principle can be delineated which is sure, clear—in a word, truly mathematical[76]—which can appropriately be substituted for "the infinite".[77] This is to be done without making the researches which had been expedited by using the concept of "the infinite" too difficult or tedious.[78] We require that this matter be treated with all possible rigor, clarity, and simplicity.[79]

Lagrange did not get what he wanted. Although the prize was awarded to Simon L'Huilier, the Berlin Academy was not satisfied with any of the many contributions received. The criticisms expressed of the contributions show that Lagrange and his fellow judges were employing higher standards than were then customary in discussions of foundations, and that they were in agreement with the criteria for rigor discussed in chapter 1. The contributions had all lacked "clarity, simplicity, and especially rigor," the committee report said.[80] Most of them had not even seen that the principle desired had to be "not limited to the infinitesimal calculus, but extended to Algebra, and to Geometry treated in the manner of the Ancients." From this it appears that most of the contributors had tried to find some ad hoc principle such as the compensation of errors, thereby avoiding the reduction of the calculus to more general mathematical concepts.

Furthermore, the contributors had not justified the wealth of received results of the calculus—which is, I have argued, absolutely essential for a real foundation: "They have all forgotten to explain how so many true theorems have been deduced from a contradictory supposition." The tone of the committee report, especially its statement that the prize question "had received no complete answer," shows that L'Huilier's essay was regarded as the best of a bad lot.[81] Lagrange must have been even more dissatisfied with the state of the foundations of the calculus after having read these essays.

But dissatisfaction with an old theory is not in itself enough to make a man create a new one. Lagrange did not convert his dissatisfaction into a new treatment of the problem.[82] At this time, he was completing his *Mécanique analytique*, not working on the foundations of analysis. But the *Mécanique analytique* may have provided Lagrange with one more reason to want to make calculus rigorous; now that, as he believed, he had reduced all of mechanics to calculus, the calculus itself needed to be made rigorous in order to make mechanics rigorous.[83] Nevertheless, when he moved from Berlin to Paris in 1786, he fell into a long period of depression during which he did no work at all.[84]

The New Departure: Lagrange's Theory of Analytic Functions

Had it not been for the French Revolution, after which Lagrange was pressed into service to teach analysis at the newly founded Ecole Polytechnique, he might never have written his *Fonctions analytiques*. The mathematical world would have had to wait for somebody else to be the first to publish a full-scale work principally devoted to establishing the calculus and all its results on an algebraic foundation. Lagrange may well have had many of the ideas of his *Fonctions analytiques* before he began to teach at the Ecole Polytechnique, but it was the need to teach there which led him to put the ideas together and make them public; as he himself put it, he had been "engaged by particular circumstances to develop the general principles of analysis."[85] For the task of teaching the calculus, Lagrange felt that it was no longer enough simply to recognize that infinitesimals, limits, and prime and ultimate ratios were inadequate foundations; a positive doctrine was needed. Accordingly, he recalled his "old ideas" on the principles of the differential calculus and worked them out further. The "old ideas," first expressed in 1772, appear clearly in the full title of Lagrange's book of 1797: "Theory of analytic functions, detached from any consideration of infinitely small or evanescent quantities, of limits or of fluxions, and *reduced to the algebraic analysis* [*analyse algébrique*] *of finite quantities*" [italics mine]. The last phrase suggests the origin of Cauchy's program of reducing the calculus to "analyse algébrique."

We may distinguish several reasons for Lagrange's innovation of 1797. First, there was the immediate cause: the need to teach a course in analysis. Second, there were the demands Lagrange saw arising from the recent history of mathematics. His lifetime of reflection on the inadequacies of previous foundations of the calculus convinced him that a new basis was needed, and he felt a real need to synthesize the results of the past century. Finally, he had a method—the algebra of infinite series—which was part of the algebraic tradition he so prized and which he had already sketched as a foundation for the calculus as long ago as 1772.

Lagrange began his *Fonctions analytiques* with a definitive critique of all previous foundations of calculus, and wrote a brief article adding to those criticisms in 1799.[86] Lagrange's critique was not original in its particulars; its originality lay, rather, in the use he made of it—as a preface to a new approach to the foundations of the calculus, in which both advanced and elementary results would be exhibited as consequences of the definitions. Thus his critique, unlike earlier ones, became part of the standard literature of analysis, and Cauchy, Bolzano, and their contemporaries could not ignore it.

Infinitesimals were not rigorous at all, in Lagrange's view. Leibniz, L'Hôpital, and Bernoulli, "content with reaching exact results . . . in a prompt and sure way . . . did not occupy themselves with demonstrating the principles" of the calculus.[87] The infinitesimal calculus got exact results only by the compensation of errors; and unfortunately this fact could not be used as a foundation for the calculus, because "it would be difficult to give a demonstration" that the errors are always compensated.[88]

Newton's fluxions were unacceptable because they consider mathematical quantities as if "engendered by motion." To be sure, the view has a certain deceptive plausibility, since "everyone has or believes to have an idea of velocity." But we do *not* have a clear enough idea of an instantaneous variable velocity.[89] A more fundamental objection, in Lagrange's view, is that the calculus is mathematics, not physics; it should have "only algebraic quantities as its object." Velocity is thus a "foreign idea"; Newton's fluxionary explanation is therefore untenable.[90]

As for the limit concept as it then existed, Lagrange held that it was vague, too narrow, and geometric rather than algebraic. Lagrange's argument for the vagueness of the concept of the limit of a ratio whose terms go to zero is like Berkeley's, though he cited no source: the limit concept considers quantities "in the state in which they cease, so to speak, to be quantities." The ratio of two finite quantities "no longer offers a clear and precise idea to the mind, when the terms of the ratio become zero simultaneously."[91] To show that the limit concept was not broad enough to serve as a foundation for the calculus, Lagrange observed that if a quantity could never surpass its limit, then the subtangent could not always be defined as the limit of a subsecant, since "nothing prevents the subsecant from increasing still when it has become the subtangent."[92] Finally, the idea of limit is based on the example of a curve as a limit of a sequence of polygons;[93] thus it is essentially a geometric idea and foreign to the spirit of analysis.[94] For Lagrange, the calculus was essentially algebraic and therefore need not—and should not—be founded on principles borrowed from other subjects.

Lagrange had one more criticism of the existing foundations. Although he did not state this criticism explicitly in 1797, the entire structure of *Fonctions analytiques* implies it: the older foundations were used merely as introductory matter; they could not support all the received results. Accordingly, *Fonctions analytiques* differed from its predecessors in a crucial respect: It was intended primarily to establish the rigor of the calculus rather than to derive new results; its definitions were not relegated to an introduction, never to be seen again. The book certainly was not indifferent to new results, as the example of the Lagrange remainder for the Taylor series should remind us. But its outstanding feature is that it does full justice to the already existing wealth of results in the calculus. Lagrange deduced known results of great complexity from his new foundation, not only in the calculus but in geometry and mechanics as well.

It was the force of Lagrange's example that did most to change the prevailing attitudes toward foundations, through his teaching at the Ecole Polytechnique and through his widely read and highly influential textbooks.[95] Both Bolzano and Cauchy were influenced by Lagrange's views. Bolzano cited Lagrange's work in the paper in which he echoed Lagrange's call for abandoning appeals to geometry and motion in favor of algebra as the very title of his "*purely analytical proof* of the theorem that between any two values which give results of opposite sign there is at least one real root of the equation" [italics mine] shows.[96] When Cauchy went on his first major engineering job in 1810, his biographer reports, the *Fonctions analytiques* was one of the four books he took with him.[97]

Lagrange had no idea of the extreme fruitfulness that was to result from the new foundations of the calculus; he did not anticipate the general theory of convergence, limits, and the definite integral. Nevertheless, in recognizing the vital importance of the question of rigor, he took a major step in the revolution which we associate with the name of Cauchy: changing the way mathematics was done so that the foundations of the calculus became essential, not peripheral. From Lagrange's work, Cauchy and Bolzano learned not just techniques, but an attitude toward foundations different from the one prevailing. This new attitude was essential if the calculus were to be given a firm foundation.

However, it is fortunate that eighteenth-century mathematicians did not concentrate on foundations to the exclusion of other questions. Through their successful concern with particular problems, these analysts developed the mathematical substance for which Cauchy could provide the rigorous basis; and while they were solving problems, they also evolved many of the concepts and techniques which Cauchy needed. This combination of techniques and results with Lagrange's program for reducing the calculus to algebra provided the necessary conditions for Cauchy's achievement.

3 The Algebraic Background of Cauchy's New Analysis

Introduction

By the end of the eighteenth century, many mathematicians realized the need to make the calculus rigorous. But merely wanting to solve a problem does not in itself provide the solution. Nor were contemporary discussions of the foundations of the calculus of much help in providing the necessary rigor. The contrast in sophistication between Cauchy's successful rigorous proofs and earlier arguments about limits and fluxions is overwhelming. Yet, however superior Cauchy's proofs may have been, it is my thesis that he owed much to the past. On what, then, did Cauchy build?

Cauchy used, transformed, and extended a number of ideas from various branches of eighteenth-century mathematics. As an example, Cauchy took a number of particular results in analysis—for instance, that certain infinite series had finite sums—and rendered them at once both rigorous and general. Some of his theory of derivatives arose from proofs about derivatives given by Lagrange and Ampère. Cauchy was led to some important theorems about integration by earlier approximations to the values of definite integrals and to the solutions of differential equations. But the source of many of the general ideas and techniques of Cauchy's rigorous calculus is algebra, especially the algebra of approximation.

Cauchy used Lagrange's phrase "analyse algébrique" in describing the content of the *Cours d'analyse*; it is thus taking Cauchy at his own word to look for the origins of his ideas and techniques in algebra. The dependence of nineteenth-century calculus on eighteenth-century algebra has been given little attention by historians, however. This neglect may be due partly to the modern habit of dividing mathematics into specialties, and to the corresponding assumption that if the calculus is one subject, it must have one history. But although specialization in mathematics has seemed necessary in the complex mathematical world of the nineteenth and twentieth centuries, it was foreign to the spirit of eighteenth-century mathematics. One cannot call Newton, Euler, or Lagrange algebraists or analysts; they were proficient in all areas of mathematics. Cauchy did not learn calculus and algebra as two entirely disconnected subjects. He was not restricted to works explicitly given over to discussions of limits and fluxions in looking for techniques to make the calculus rigorous. Cauchy's "analyse algébrique" is a grand synthesis of algebraic methods with the basic concepts of analysis.

Both the theory and the practice of algebra contributed to the rigorization of analysis. Views on the theory of algebra led some eighteenth-century mathematicians, most notably Lagrange, to think that algebra was rigorous and, therefore, that the calculus could be made rigorous by being reduced to algebra. Also, many mathematicians of the period viewed infinite processes as a part of algebra, a belief which makes infinite processes belong equally to algebra and to analysis. Infinite processes were the basis of the first attempt to reduce the calculus to algebra, an attempt which, though unsuccessful, was important in a number of ways.

Nevertheless, from a modern point of view, the algebraic basis of rigorous calculus is not the so-called algebra of infinite processes; the basic techniques come instead from the algebra of inequalities. By what may seem a fortunate coincidence, in Cauchy's time the algebra of inequalities had matured into a set of useful, applicable, well-developed techniques, and was thus ready and at hand for his use in founding the calculus. These inequality techniques came primarily from eighteenth-century work on approximating the solution of algebraic equations.

Obviously the study of approximations requires some use of inequalities. Besides providing some technical facility in manipulating inequalities, approximations were valuable to Cauchy in two main ways. First, the study of error bounds in approximations led to work on specific types of inequalities much like those needed in the study of convergence. Second, the study of approximating processes provided ways of constructing quantities—constructions that could be converted into existence proofs. An illustrative example will be given to show the way in which an algebraic approximation provided

Cauchy with the inequality techniques he needed to prove a major theorem in analysis. This example—Cauchy's proof of the intermediate-value theorem for continuous functions—provides not only a striking piece of evidence for the algebraic origins of rigor in analysis and a specific instance of the conversion of an approximation into a proof of the existence of the intermediate value, but also a detailed illustration of the relation of Cauchy's innovations to the work of his predecessors.

The Theory of Algebra: The Certainty of Universal Arithmetic

The calculus was to be made rigorous. How could this be done? Its first principles had to have the certainty, generality, and self-sufficiency expected of a rigorous subject. Since Greek geometry was viewed as too cumbersome and too foreign to the spirit of the calculus to serve as its foundation, there was only one subject available with the desired characteristics: algebra.

Eighteenth-century mathematicians based their faith in the generality and certainty of algebra on the view that algebra was a "universal arithmetic."[1] In universal arithmetic, the operations of ordinary arithmetic were applied to letters; the letters were understood to represent any numbers whatsoever. Thus mathematicians could obtain complicated symbolic relations, which yielded valid arithmetical results when numbers were substituted for the letters.

Although the statements of "universal arithmetic" were only as valid as the laws of ordinary arithmetic, this was saying a good deal. Usually in the eighteenth century (and even in the nineteenth), arithmetic was considered well founded and since algebra was just a generalized arithmetic, the truth of its conclusions was believed as well founded, as the truth of arithmetic.[2] The idea of universal arithmetic thus accounts for the widespread eighteenth-century belief in the certainty of algebra, a belief which was necessary in order to convince people that the calculus could be made rigorous by being reduced to algebra.

Besides stressing generality and certainty, the description of algebra as universal arithmetic called attention to algebraic symbolism, and therefore to the heuristic power of symbolic notation. Powerful notation was, of course, not unique to algebra. Leibniz had provided a highly useful notation for the calculus, and his notation gave much of eighteenth-century calculus a sort of algebraic style. Practitioners of the calculus thus naturally saw algebra as a kindred subject, a view which made it seem plausible that the already established certainty of algebra could somehow be carried over into the calculus.

Another factor making algebra seem an appropriate foundation for the calculus, especially to Lagrange, was that seventeenth- and eighteenth-century mathematicians considered infinite series and other infinite expressions part of algebra. The study of infinite series was a major area of algebraic activity in the seventeenth century and was among the major concerns of Newton and Leibniz. It was thought obvious that algebraic methods could routinely be extended beyond the finite operations of ordinary arithmetic to infinite processes. For instance, repeated additions produced infinite series, and so did dividing one polynomial by another of higher degree. Repeated multiplications and divisions produced infinite products and continued fractions. The philosophy of algebra included, therefore, infinite as well as finite processes.

The extensions of algebraic methods from the finite to the infinite domain could be justified in terms of the idea of universal arithmetic. The desire for the greatest possible generality led to the hope that almost any operation which could be carried out meaningfully in the finite domain could also be carried out in the infinite one.[3] The universal acceptance of infinite decimal expansions along with finite arithmetic provided a ready analogy for the jump from finite to infinite algebraic operations.[4] In addition, belief in the certainty of universal arithmetic led in practice to an excessive reliance on symbolism and to the assumption that formal validity carried a built-in guarantee of truth. Faith in the power of notation helped encourage mathematicians to apply the same sorts of techniques to infinite processes that they had applied to finite processes. Throughout the eighteenth century, infinite power series were added, multiplied, and converted into infinite products, just as if they had been very long polynomials. These procedures may not have been rigorously established, but they were considered justified by the wealth of results they yielded in applications to equation solving, geometry, and the calculus. From these attitudes toward infinite processes, the conception of the algebra or analysis of the infinite was a natural development.

The study of infinite processes may be said to have come of age in 1748, when Euler published his *Introductio in analysin infinitorum*, a work which studied infinite series, infinite products, and infinite continued fractions. The *Introductio* sought to give an account of infinite "analytic expressions," just as theories of equations had given an account of finite ones.[5] Among many other achievements in this book, Euler gave infinite-series developments for all the standard functions of the time, such as quotients of polynomials, exponentials, logarithms, sines, and cosines.[6] Many contemporary mathematicians believed that Euler had shown that the functions commonly studied in the calculus, even those usually defined geometrically, could be represented by infinite series—apparently with no appeal having been made to the concepts of the differential or the integral.[7] Deriving these series was viewed as part of algebra: the algebra or analysis of the infinite.

Lagrange and the Algebra of the Infinite

It was the idea that there was an algebra of infinite power series which led Lagrange to originate and promulgate the program of reducing the calculus to algebra—the program adopted and successfully carried out by Cauchy, Bolzano, and Weierstrass. What Euler had managed to do for particular functions in the *Introductio* convinced Lagrange that *any* function could be given an infinite power-series expansion. Since Lagrange shared the view that infinite processes were part of algebra, Euler's work probably sufficed to convince him that the calculus could be reduced to algebra.

In defining the task of the *Introductio*, Euler had said that he was studying functions, and defined a function of a variable quantity as "an analytic expression composed, in any manner, of that same quantity and of numbers or of constant quantities."[8] Lagrange, following Euler, took a function to be any "expression de calcul" into which the variable entered in any way.[9] Lagrange believed that the "algebra of series" provided any such function f with a power-series expansion

$$f(x+h) = f(x) + hp(x) + h^2q(x) + \cdots.$$

He even gave a proof—of course, fallacious—that such a power series expansion always exists except for particular isolated values of x. (Lagrange recognized that there might be a few functions without such developments;[10] but he did not think such functions could be studied by the methods of his calculus.) Lagrange then defined $p(x)$, the coefficient of h in the power series expansion of $f(x+h)$, to be the derivative $f'(x)$. Lagrange identified his $f'(x)$ with $df(x)/dx$, and used his definition of $f'(x)$ to derive the received results of the calculus.[11] For Lagrange, all the applications of the calculus, whether to algebra, geometry, or mechanics, rested on those properties of functions which could be learned by studying their Taylor-series developments.[12]

Foreshadowings of his conclusion that the calculus should be reduced to algebra, with no appeal either to geometry or to the idea of motion, may be found throughout the eighteenth century, making Lagrange's views appear to be the culmination of a long and successful tradition. D'Alembert had attacked Newton's doctrine of fluxions for using the idea of motion, since the calculus had "only algebraic quantities as its object."[13] Euler had claimed that his treatise on the differential calculus "remains throughout within the bounds of pure analysis," since it required no diagrams for its explanations.[14] Lagrange repeated many remarks of this sort; besides the ones we have already cited, there is his well-known boast that no diagrams could be found in his *Mécanique analytique*. But Lagrange was more insistent upon his goal than were d'Alembert and Euler and much more single-minded and self-consciously algebraic about working out all the implications of his algebraic foundation for the calculus. And Lagrange's work helped convince many mathematicians, especially Bolzano and Cauchy, that the calculus should be made rigorous by being reduced to algebra.

Bolzano spoke of the need to free analysis from geometry and to seek out "purely analytic proofs" for theorems of analysis. He wrote that it would be "an unendurable offense against good method to derive truths of pure (or general) mathematics (arithmetic, algebra, or analysis) from considerations which belong to purely applied (or special) parts of mathematics, such as geometry."[15] He added that "the concept of *Time* and even more that of *Motion* are just as foreign to general mathematics as that of *space* is."[16]

Cauchy adopted Lagrange's phrase "analyse algébrique" for the subtitle of his *Cours d'analyse* of 1821. And Cauchy actually carried out the reduction of the calculus to the "algebraic analysis of finite quantities" referred to in the full title of Lagrange's *Fonctions analytiques*.

Though Lagrange's Taylor-series method of reducing the calculus to algebra was not the method ultimately adopted by Cauchy, it led Lagrange directly to several insights which were important to nineteenth-century analysts. Lagrange's stress on the functional nature of the derivative helped establish that the calculus studies functions, not infinitesimal differences or geometric objects. Because of his reliance on series, the function concept he adopted was that of Euler's *Introductio*; Lagrange even called the calculus the "theory of analytic functions"—analytic functions for him being finite or infinite algebraic expressions containing real variables. Cauchy accepted Lagrange's view of the primacy of the function concept in the calculus, while returning to the conception first advanced by Euler that a function is any general dependence relation.[17]

Since Lagrange believed that the Taylor-series development of a function $f(x)$ was essentially an algebraic process, he came to treat derivatives of *all* orders as functions, not ratios of infinitesimals. He defined the first derived function $f'(x)$ as the coefficient of h in the Taylor-series expansion of $f(x+h)$ and, recursively, $f''(x)$ as the first derived function of $f'(x)$—that is, as $[f'(x)]'$—and so on. Thus $f^{(k)}(x)$ was, for Lagrange, the same sort of function as $f^{(k-1)}(x)$. The value of this insight is that it is only when we stop thinking of $f'(x)$ as some sort of quotient, and only when we stop thinking of d^2y/dx^2 as a different sort of object (since it is a quotient of second-order, rather than first-order, infinitesimals) from dy/dx, that we can be sure of getting away from what Lagrange called "the false idea of the infinitely small."[18] Lagrange's notation $f'(x)$, his recursive definition of the successive derived functions, and his term "fonction derivée" soon caught on and aided a clearer understanding of the concepts of the calculus.

Cauchy was strongly influenced by Lagrange's view that the calculus should be reduced to algebra. In addition, he adopted several of Lagrange's algebraically induced innovations. Nevertheless, Cauchy did not accept the particular algebraic foundation used by Lagrange. He found it lacking in both rigor and generality. Lagrange had allowed almost all the methods used in the algebra of the infinite to be brought into the calculus.[19] Cauchy, however, had well-founded doubts about the automatic general interpretation of symbolic expressions. He had warned that "most [algebraic] formulas hold true only under certain conditions, and for certain values of the quantities they contain."[20] In particular, relations about infinite series hold only when the series are convergent.[21] This was, I believe, Cauchy's principal reason for concluding that the calculus cannot be founded on the algebra of power series. In addition, Cauchy observed that different functions can sometimes have the same Taylor series.[22] Politely but firmly, and "in spite of all the respect that such a great authority must command," Cauchy rejected Lagrange's foundation for the calculus.

Cauchy, unlike Lagrange, came to see that it was the algebra of inequalities, not of equalities, which could provide a basis for the calculus. The search for the origins of Cauchy's innovations must turn, therefore, to the origin and development of inequality techniques in eighteenth-century algebraic practice—in particular, in the treatment of approximations.

Eighteenth-Century Algebra: Theory and Practice

To use the algebra of finite quantities as a basis for the calculus, it was not enough for algebra to be rigorous and to use powerful notation. Algebra also had to provide techniques appropriate for proofs in analysis. These would have to come, not from the philosophy of algebra but from the practice of algebraists. Fortunately for this purpose, the practice of algebraists in the eighteenth century was different from what the idea of universal arithmetic would suggest. No algebraist of the period regarded the purpose of algebra as the mere proclaiming of general rules for finite (or infinite) arithmetical operations, even in symbolic language.[23] If algebra was in theory a universal arithmetic, it was in practice an "analytic art": the art of solving equations.[24]

To be sure, there was a "theory of equations," which could be included under the description universal arithmetic. But the theory of equations, which gave the relations between the roots of a polynomial equation and the coefficients of the powers of the unknown, was treated chiefly as a means to the end of equation solving.[25] Almost any method suitable to that end was considered an acceptable part of algebra. Equations were to be solved exactly when possible, but an approximate result would be better than none.

Accordingly, many methods of approximating the solutions of algebraic equations were devised. The need for these methods was obvious. Methods of solving algebraic equations exactly were available only for equations of the first, second, third, and fourth degrees; for a method of solution to be applicable to any polynomial equation, the method would have to be one of approximation. As an added advantage, approximation methods could often be applied to transcendental as well as polynomial equations. Approximation techniques were "general" in a sense in which exact methods of solution were not.

These techniques led naturally to the study of the algebra of inequalities. Thus the major motivation for that study was not theoretical, but practical: solving equations. Nevertheless, the algebra of inequalities raised no philosophical problem for theorists. Since the algebra of inequalities originated as a shorthand way of expressing simple relations of order between magnitudes, it could easily be considered part of universal arithmetic. Thus inequalities, like infinite series, were comfortably at home in both the theory and practice of eighteenth-century algebra.

The algebra of approximations was a source of rigorous calculus in several ways. First, the use of inequalities in approximation techniques, especially in error estimates, helped develop skill in manipulating algebraic inequalities and made such manipulations available for use in convergence and delta-epsilon proofs by Cauchy and his successors. Second, an approximation technique supplied with a precisely computable error bound can be converted by someone like Cauchy, in possession of the modern definition of convergence, into a proof that the approximation converges. Third, a sequence of approximations to a quantity sometimes can be converted into a proof of that quantity's existence by providing a way to construct it as a limit of the sequence. Mention should be made here that the interest in approximations in the eighteenth century went far beyond algebra. Approximations to the values of Taylor series and of definite integrals played important roles in problem solving and were used by Cauchy in his rigorous theories of the derivative and the integral. Finally, the relation in approximations between infinite processes and error estimates brought together the algebra of the infinite and the algebra of inequalities; these are precisely the two subjects whose fusion lies at the heart of the nineteenth-century rigorization of analysis.

Eighteenth-Century Algebraic Approximations

A wide variety of methods of approximation may be found in eighteenth-century algebra.[26] Some generalizations apply to almost all the methods, however. First, the goal of most approximation procedures was to derive infinite analytic expressions for the roots of any polynomial equation. If such an infinite form could not be found, the approximator would give a recursive process for approximating the root, which could then be applied as many times as desired. But the full infinite form was always preferred when such a form could be obtained. Most mathematicians preferred to write down infinite expressions, which, since they appear in equations, seemed to give the solutions *exactly*, rather than to write down inequalities.[27] Even in those rare instances when methods of approximation were evolved by inequality considerations rather than by algebraic substitutions, mathematicians preferred to convert the final result to an infinite analytic expression.* This example of the popularity of the algebra of the infinite

*Johann Heinrich Lambert, for instance, solved the equation $x^m + px = q$ by deriving a sequence of inequalities bounding x above and below that became closer and closer to x. But his final result was stated as an infinite series

serves to remind us that a developed algebra of inequalities does not arise automatically and necessarily out of work on approximations. More than the simple interest in approximations will be necessary to explain the genesis of sophisticated inequality techniques.

A second common methodological feature was the stress on particular numerical examples. Although generality was often said to be a goal, in fact particular numerical examples abounded. And although the purpose of the examples was only illustrative, their effect was to block off the consideration of highly unusual or pathological counterexamples and therefore the systematic consideration of error estimates. This orientation toward particular results parallels the period's approach to the calculus.

A third point is that until the late 1760s there were almost no explicit general error estimates, except for a few simple results like the error term for the sum of a geometric progression. Although enough information existed in some approximating formulas to provide error estimates, most mathematicians felt that they had more important things to do. Here as in the calculus, the desire for explicit results governed which questions were asked. When, for instance, Newton and Euler appealed to inequalities to justify approximations, it was usually to ensure that the terms of the specific infinite series under consideration rapidly became smaller and smaller.[29] The diminution of the first few terms of an infinite series was generally treated as sufficient to ensure that the values of the successive approximations became closer and closer to the root. Only in the latter part of the century was any serious attention paid to computing general bounds on the errors made in approximations, and not until Cauchy are there explicit proofs for the convergence of approximations.

To appreciate the relatively elementary way inequalities were used in the early 1700s, let us look at Newton's

---

whose terms alternate in sign:

$$x = a/p + q^m/p^{m+1} + mq^{2m-1}/p^{2m-1} - [m(3m-2)/2]q^{3m-2}/p^{3m+1} + \cdots.$$

He made no further mention of his generating inequalities.[28]

approximation, which was widely discussed throughout the century. Newton's exposition was not general, but was presented for a simple cubic equation from which the generalization was obvious:

$$y^3 - 2y - 5 = 0.$$[30]

As a first approximation to the solution of this equation, Newton chose 2 "as differing from the true root less than by a tenth part of itself."[31] Setting $y = 2 + p$, and substituting in the original equation, Newton obtained an auxiliary equation for p:

$$p^3 + 6p^2 + 10p - 1 = 0.$$

But since p is small, the higher-order terms can be neglected, yielding

$$10p - 1 = 0, \qquad p = 0.1.$$

[Though Newton did not point it out, this procedure is equivalent to solving a polynomial equation $P(y) = 0$, given a first approximation $y = a$, by letting the second approximation be $y = a + p = a - P(a)/P'(a)$.][32] The procedure can then be repeated as often as desired.[33]

Newton was sufficiently concerned about the accuracy of the approximation to give an inequality condition not only for the first approximation but for the neglect of the higher-order terms in the cubic equation for p. Let us write that equation in general as

$$p^3 + cp^2 + bp + e = 0.$$

Newton said that the p^2 term could be neglected in solving for p only when $10ec < b^2$. Newton gave no error estimate, nor any other explicit justification for this requirement.[34] Nevertheless Newton's inequality is instructive in showing that even in the seventeenth century, discussing the accuracy of an approximation technique involved inequalities.

As an illustration of the state of these questions in the mid-eighteenth century, consider Euler's application of Newton's method to the equation $y = x^n - a^n - b = 0$. By Newton's method, Euler's first approximation was $f = a + b/na^{n-1}$.[35] Euler then asked the question Newton had asked: How small must b be to justify neglecting the higher-order terms in the Taylor-series expansion for y? Euler's answer was

$$a^{n+b} < (a+1)^n, \quad (3.1)$$

where a and b presumably are positive. He did not explain from where he derived this condition,[36] and it is not valid.[37] The particular numerical examples used by Euler always satisfy this inequality, however, because in illustrating the use of the method, he tried it out on equations for which n is relatively small, and he usually made $f - a$ much smaller than his criterion required.[38] This type of inequality consideration is not sufficiently sophisticated to be useful in general discussions of the speed of convergence[39] or the error of an approximation, to say nothing of the question whether the approximation converges at all. For more sophisticated discussions we must turn to the work of d'Alembert and Lagrange.

Speed of Convergence and Bounds on Error

In the last third of the eighteenth century, there was a shift in interest among algebraists from merely deriving approximation procedures to deriving precise error estimates and measures of speed of convergence. This shift parallels the change in attitude toward the calculus. In both subjects, a preoccupation with results gradually led to a greater concern with precision and a desire to prove what was known. The inequalities derived for estimating speed of convergence were themselves complicated and provided facility in inequality computation; in addition, in some cases they assisted in the computation of error bounds.

The interest in error bounds arises naturally; an approximation provides the solution to an equation either as an infinite expression or as a recursive sequence of successive approximations. In either case, however, ordinary mortals can apply the process only some finite number of times. The full infinite expression is the solution to the equation, its "true value." But though the true value is known in principle, in practice the only obtainable *exact* result is the computation of the maximal possible error after some finite number of steps.

Algebraic error estimates may not seem to have any connection with the foundations of the calculus. Yet in fact they have a direct connection, for an expression bounding the error in an infinite-series approximation can be used in two directions. An eighteenth-century algebraist like d'Alembert or Lagrange often wanted to find the maximal error after the nth approximation for some given n. Cauchy, on the other hand, used these expressions in reverse to compute, given a maximal error, the value of n necessary for that accuracy; his definition of limit could then be used to show that the approximating process converged.[40]

Another link between error bounds and the calculus came from the particular turn Lagrange's mind took in applying the technique of error estimates to the Taylor series. The desire to show that the errors in a Taylor series are bounded, just like the errors in any other good approximation, led Lagrange to the Lagrange remainder of the Taylor series. He let the function $f^{(n)}(x)$ have its maximum on the interval $(x, x+h)$ at $x = q$, its minimum at $x = p$, and derived the formula

$$\begin{aligned} & f(x) + hf'(x) + \cdots + (h^n/n!)f^{(n)}(p) \\ & \leqslant f(x+h) \\ & \leqslant f(x) + hf'(x) + \cdots + (h^n/n!)f^{(n)}(q). \end{aligned} \tag{3.2}$$[41]

Although Lagrange's result (3.2) may not appear algebraic to us, it was for him a typical algebraic result, in both conception and proof. Lagrange's derivation of (3.2), which rested heavily on the algebra of inequalities, provided Cauchy with essential techniques for proving theorems about derivatives.

In 1768, d'Alembert published the fifth volume of his mathematical papers, in which appeared a long paper on diverse topics that included the question of the convergence of the binomial series.[42] The binomial series for $(1 + \mu)^m$ when m is rational, discovered by Newton, was often used as a way of approximating the solution of equations like $x^{p/q} = 1 + \mu$. D'Alembert's paper contained a careful, purposeful computation of the bound on the error made in this approximation. He used no techniques that had not been available to Euler in the latter's discussion of the series $(\mathrm{a}^n + \mathrm{b})^{1/n}$, but the questions he asked about the binomial series could be extended to more general investigations; they were to influence both Lagrange and Cauchy and thus figure in the process by which the inequality proofs used in computing bounds on error in approximations eventually were converted into rigorous proofs in analysis.

D'Alembert treated two important questions about the binomial series for $(1 + \mu)^m$. First, under what circumstances do the successive terms decrease? Second, what are the bounds on the error made in approximating the sum of the infinite series by the sum of a finite number of its terms? D'Alembert's work exemplified the power of the algebra of inequalities to solve such problems. This, and a paper by Lagrange,[43] is the earliest example of the simultaneous treatment of these two questions and the first sophisticated use of inequality techniques to find the answers. First, d'Alembert computed the n for which the absolute value of the ratio of successive terms of the series is less than unity (though he did not have a notation for absolute value, he did have the idea, consistently using the expression "abstraction being made of the sign"), thereby calling attention to one important property of convergent infinite series—thus the name *d'Alembert's ratio test.** Second and more important, d'Alembert computed the bounds on the error made in taking a finite number of terms in the series for $(1 + \mu)^m$ instead of the sum of the infinite series.

Consider d'Alembert's search for the n such that a particular "convergence" inequality holds. It is easy to compute the ratio of the successive terms in the binomial series expansion for $(1 + \mu)^m$; the ratio between the $(n + 1)$st and nth terms is given by d'Alembert as $\mu(m - n + 1)/n$. Now consider with d'Alembert the binomial expansion for $(1 + 200/199)^{1/2}$.[44] The first few terms—even the first 100 terms—certainly seem to get successively smaller; it appears to be a perfectly good approximating series. But by applying his ratio test, d'Alembert found that this was not so. The general ratio of the $(n + 1)$st term to the nth term in this example has as its

* His use of the term *converge*, together with the explicit computation of the ratio, has earned d'Alembert a bit of undeserved credit for showing *in general* that a series that satisfies the ratio test converges. But, as I document at length in chapter 4, d'Alembert—and several of his contemporaries—meant by converge nothing more than the successive decreasing, in absolute value, of the terms.

absolute value $(1 + 199)(1 - 3/2n)$. Can this ratio ever exceed unity? Yes, said d'Alembert, and he computed that the ratio is greater than 1 for all n greater than 300. "Thus," he concluded, "it is wrong to believe that a series is truly convergent [i.e., in d'Alembert's usage of the term, the terms continually diminish] because it converges, even very strongly, in its first terms." His discussion is really a warning against the widely, though implicitly, used assumption that it was enough to look at the first terms of an infinite approximating series—as Euler, for instance, sometimes had done—to predict the behavior of the infinite series. Furthermore, d'Alembert had given one of the first examples of a vitally important technique: computing for which n a particular "convergence" inequality holds.[45]

D'Alembert then stated that the best situation in working in infinite series was that in which all the terms decrease after the first and all the terms have the same sign.[46] When these conditions hold for the binomial series, d'Alembert was able to compute the bounds on the "error" —that is, the difference between the nth partial sum and the sum of the infinite series.[47] Since the ratio between the $(n + 1)$st and nth terms of $(1 + \mu)^m$ is $\mu[1 - (m + 1)/n]$, d'Alembert explained his result as follows:

If the terms of the series have the same sign beginning with some n, so that $n > 1 + m$, it is easy to see that the sum of the series, beginning with the nth term which I designate by A, is

$$< A + A\mu + A\mu^2 + A\mu^3 \text{ \&c.},$$

and, on the other hand,

$$> A + A\mu(1 - m + 1/n) + A\mu^2(1 - m + 1/n)^2 + A\mu^3(1 - m + 1/n)^3 + \text{\&c.}$$

Thus the sum of the terms from A on will be $< A/1 - \mu$ and $> A/(1 - \mu(1 - m + 1/n))$ . . . which gives a practicable enough way of finding the sum of the series by approximation. The error will be less than

$$A/1 - \mu - A/[1 - \mu(1 - m + 1/n)].^{48}$$

D'Alembert did not explain why the sum of the series from the nth term on was included between those two geometric progressions, and careful attention to absolute values is needed to make his result correct. If we take absolute values throughout, and let S be the sum of the series, S_n the $(n-1)$st partial sum, d'Alembert's result becomes

$$(3.3)\quad |A| + |A\mu(1 - m + 1/n)| + |A\mu^2(1 - m + 1/n)^2| + |A\mu^3(1 - m + 1/n)^3| + \cdots < |S - S_n| < |A| + |A\mu| + |A\mu^2| + |A\mu^3| + \cdots.$$

Now indeed "it is easy to see' (3.3) by comparing the terms of the binomial series with the two bounding geometric progressions* and recalling that all terms beyond the nth have the same sign. And d'Alembert showed that he knew in practice how to deal with the absolute values—for instance, that $|S - S_n| < |A/1 - |u||$—in working out examples.[49]

Though d'Alembert gave this error-bound computation for the binomial series only, it is clear that the points he raised have broad implications. D'Alembert had given a completely worked-out example of how the partial sums of a series could be proved to differ as closely as desired from some fixed value. To be sure, it had never occurred to him to question the existence of the sum of the series. Nevertheless, in Cauchy's terms, d'Alembert's result can easily be converted into a proof of the convergence of the series. D'Alembert's argument rests on a term-by-term comparison with a geometric progression. And the idea of proving that an arbitrary series converges by comparing the series with a convergent geometric progression was essential to Cauchy's derivation of the root test for convergence. Cauchy may have known d'Alembert's paper directly and surely knew it through its summary in S. F. Lacroix's *Traité du calcul*.[50]

Computations analogous to d'Alembert's were later undertaken for other approximations. Even before having read d'Alembert's paper, Lagrange gave an error estimate

* First observe that since he assumed $n > m + 1$, $(1 - m + 1/n) < 1$. Now for the right-hand part of (3.3), for the $(n+1)$st term $|A(1 - m + 1/n)\mu| < |A\mu|$. Similarly, for the next term of the binomial series, $|A(1 - m + 1/n) \cdot (1 - m + 1/n + 1)\mu^2| < A\mu^2$, and so on. For the left-hand part of (3.3), note first that $(1 - m + 1/n) < (1 - m + 1/n + 1)$. Then $|A\mu^2(1 - m + 1/n)^2| < |A\mu^2(1 - m + 1/n) \cdot (1 - m + 1/n + 1)|$, and so on for higher-order terms.

for an approximation of his own that used continued fractions to solve algebraic equations.[51] Together with d'Alembert's 1768 paper, Lagrange's approximation and its error bound constitute the best of the theory of algebraic error estimates until nearly the end of the century.

General Error Estimates: Lagrange's *Equations numériques*

In 1798 Lagrange published a work almost entirely devoted to approximation techniques in algebra, the first such organized treatment of the subject: *Traité de la résolution des équations numériques de tous les degrés*.[52] Some of the approximations he discussed were his own, others not, but he approached all the methods in the same spirit. Of course he, like his predecessors, wanted the approximations to converge, preferably rapidly. But Lagrange did not follow Euler in letting the acceptability of an approximation rest on nothing more than the accident of convergence for some particular set of numbers. When he could, Lagrange computed the bounds on the error explicitly.[53] When he could not compute the bounds, he tried at least to show the conditions under which a second approximation was closer than the first.[54] If neither of these tasks could be accomplished, Lagrange at least described the conditions under which an approximation method did not work at all.[55]

The *Equations numériques* for the first time presented the study of algebraic approximations and the corresponding inequality techniques as a coherent subject. After Lagrange's work, there existed a systematic treatment of algebraic approximations, based on intricate manipulations of algebraic inequalities. This synthesis made available a body of inequality techniques that could be used not only to treat the convergence of approximations, but also to provide techniques for rigorous proofs in analysis.

For example, Lagrange gave the first extensive investigation of the convergence of Newton's method. Lagrange used his facility in manipulating inequalities to treat a fundamental question that he raised about many approximation methods: Under precisely what conditions do the successive approximations get closer to the root? Using the questions (and terminology) employed by d'Alembert in 1768 in discussing the binomial series, Lagrange observed that the Newton approximation might "converge" very slowly, or might even "diverge after having been convergent." And, in the spirit of the ratio test pioneered by d'Alembert, Lagrange computed the absolute value of the ratios of the successive errors in the approximation (saying, as had d'Alembert, "abstraction being made of the sign") to see under what circumstances they decreased.[56]

Lagrange, like Euler but unlike Newton, presented a general discussion of Newton's method in the notation of the calculus. In Lagrange's notation, by Newton's method, if the equation to be solved is written $F(x) = 0$, and if a is the first approximation to the root, set $x = a + p$ and then substitute this into $F(x) = 0$ to obtain

$$F(a) + pF'(a) + (p^2/2)F''(a) + \cdots = 0.$$

Neglecting the higher-order terms leads to Newton's result, $p = -F(a)/F'(a)$.[57]

Lagrange pointed out that Newton's method proceeds by applying this technique again and again. The validity of the method, then, requires that the second approximation $a + p$ be closer to the root than the first approximation a. Lagrange found a necessary and sufficient condition for the second approximation to be closer than the first, the inequality (3.5). He began his derivation of that condition by letting the polynomial to be solved be

$$x^m - Ax^{m-1} + Bx^{m-2} - \cdots = 0.$$

He supposed α to be the desired root, a the first approximation, and $a + p$ the second approximation; the other roots of the equation are β, γ, etc. For the second approximation to be closer than the first,

$$|\alpha - (a + p)| < |\alpha - a|,$$

or, equivalently,

$$\left|\frac{1}{\alpha - a - p}\right| > \left|\frac{1}{\alpha - a}\right|. \tag{3.4}$$

(He did not use absolute value notation, saying "abstraction being made of the sign" instead.) Writing the original polynomial as a product of linear factors $(x - \alpha)(x - \beta) \cdot (x - \gamma) \cdots = 0$, and defining R by $R = 1/(\beta - a) + 1/(\gamma - a) + \cdots$, Lagrange showed (3.4) to be equivalent to

$$2(\alpha - a)R + 1 > 0. \tag{3.5}$$

Here is Lagrange's derivation of (3.5). Since $p = -F(a)/F'(a)$, and since $F(x) = (x - \alpha)(x - \beta) \cdot (x - \gamma) \cdots$,

$$F'(x) = (x - \beta)(x - \gamma) \cdots + (x - \alpha)(x - \gamma) \cdots + (x - \alpha)(x - \beta) + \cdots$$

and

$$p = -F(a)/F'(a) = 1/[1/(\alpha - a) + 1/(\beta - a) + 1/(\gamma - a) + \cdots].$$

It now follows from the definition of R that

$$p = \frac{1}{1/(\alpha - a) + R}.$$

Then

$$\alpha - a - p = \alpha - a - \frac{1}{[1/(\alpha - a)] + R} = \frac{R(\alpha - a)}{[1/(\alpha - a)] + R}.$$

Therefore

$$\frac{1}{\alpha - a - p} = \frac{[1/(\alpha - a)] + R}{R - \alpha} = \frac{1}{\alpha - a} + \frac{1}{(\alpha - a)^2 R}. \tag{3.6}$$

From this it follows that if R has the same sign as $(\alpha - a)$, $(\alpha - a - p)$ will have the same sign and "the condition in question necessarily holds." But if $\alpha - a$ and R have opposite sign, then for the condition to hold, "abstraction made of signs," it is necessary that

$$\frac{1}{(\alpha - a - p)^2} > \frac{1}{(\alpha - a)^2}.$$

From equation (3.6) Lagrange obtained

$$\frac{1}{(\alpha - a - p)^2} = \frac{1}{(\alpha - a)^2} + \frac{2}{(\alpha - a)^3 R} + \frac{1}{(\alpha - a)^4 R^2}.$$

Thus

$$\frac{2}{(\alpha - a)^3 R} + \frac{1}{(\alpha - a)^4 R^2}$$

must be positive and condition (3.5) follows.[58]

Condition (3.5) is not generally useful in practice, as Lagrange pointed out, since applying it requires knowing not only α, but also R. But if the quantity a is either smaller or greater than any of the roots—a condition which can be checked without knowing all the roots—then R will have the same sign as $(\alpha - a)$.[59] Thus the method is useful for those cases.[60]

Lagrange illustrated his condition by applying it to Newton's own example, $x^3 - 2x - 5 = 0$. When $a = 2$ and the next approximation is computed, Lagrange substituted the relevant numbers into (3.5) and found that the left-hand side was $-0.1244 + 1$. Since this is positive, the second approximation must be closer than the first. But Lagrange emphasized not so much that the process converges in this particular numerical case, but *why* it does so: "In this [numerical] case, the series is, as is obvious, very convergent. We can in fact assure *a priori*, by what we have just proved, that this must be so."[61] Lagrange thus went beyond showing the goodness of approximations in particular numerical examples. He saw the convergence—that is, in his and d'Alembert's use of the term, the successive closeness—of the approximations as something that could be studied in general. The inequality computations needed in Lagrange's discussion were more intricate than those needed by d'Alembert in 1768, and Lagrange's mastery of the problems of absolute value is far superior.

We should not claim too much for Lagrange's discussion. He only gave conditions for the second approximation to be closer than the first. Though the method he used could be extended to apply to *specific* later approximations, he did not give conditions under which *all* the successive approximations decrease. And Lagrange did not show the conditions under which the error in Newton's approximation could be made less than any given quantity. Still, Lagrange's work on Newton's method was one of the first examples in which the precision of an approximation was studied by means of a careful, complicated, and purposeful argument involving inequalities.

Cauchy took up where Lagrange left off.[62] In a long note to the *Cours d'analyse* on the solution of algebraic equations by approximation, Cauchy showed himself a master of the algebra of inequalities and the use of inequalities in studying the precision of approximations. In particular, Cauchy attacked the problem of Newton's approximation, answering both Lagrange's questions and some of his own. Taking ξ to be the first approximation to the root of $F(x) = 0$, and writing $\alpha = -F(\xi)/F'(\xi)$, Cauchy worked out a set of inequalities, different from Lagrange's and much too intricate to reproduce here, too ensure that the second approximation $(\xi + \alpha)$ is closer to the root a than was the first approximation ξ. He also worked out conditions so that if $|a - \xi| < (1/10)^n$, then $|a - (\xi + \alpha)| < (1/10)^{2n}$.[63] It is clear from this condition that not only can the error be made less than any given quantity but the error decreases quite rapidly. Newton would have appreciated this condition, since it gave the precision of his approximation by comparison with decimal fractions, a type of comparison he like to make. The desirability of such a result, Cauchy noted, had been pointed out by Joseph Fourier.[64]

In illustrating the use of his error bound, Cauchy followed Lagrange in applying it to Newton's example $x^3 - 2x - 5 = 0$.[65] Cauchy's discussion is clearly in the spirit of Lagrange's work and provides a link between Lagrange's algebra and Cauchy's analysis, thereby supplying direct evidence that eighteenth-century algebraic practice helped Cauchy to feel at home with highly complicated manipulations of inequalities. Cauchy needed such complicated inequalities in his theory of convergence of series.

Of course Lagrange and d'Alembert did not compute error bounds or give an algorithm for approximating the value of a quantity in order to invent techniques that somebody like Cauchy could use to rigorize analysis. Even today, error bounds are considered worthy of study in their own right. Nevertheless, eighteenth-century approximation techniques are important because of their byproducts. In particular, their study produced a developed algebra of inequalities. And the study of approximations with error bounds made clear that inequalities could be used not only to compute speed of convergence, but—given Cauchy's new definition of convergence—to prove the very fact of convergence.

Approximations and Existence Proofs: The Origins of Cauchy's Proof of the Intermediate-Value Theorem

A method of approximation can provide a way of constructing a quantity and thus of proving its existence. Cauchy transformed previously known approximations to prove the existence of the definite integral and the solution of a differential equation. The revolution Cauchy brought about in the calculus proceeded in part by means of a transformation of techniques: he took something devised for one purpose and enlarged it so that it could be used for an entirely different purpose. Before Cauchy, approximation was viewed as a method of getting closer and closer to a real number whose existence was taken for granted. After Cauchy, real numbers were defined as limits of convergent approximating processes and their existence proved by the convergence of the approximations (given the completeness of the real numbers). One of the earliest examples of the change from the old to the new views is found in the history of Cauchy's proof of the intermediate-value theorem; an eighteenth-century approximation method led Cauchy directly to his rigorous proof of this theorem.

Cauchy shares with Bolzano the honor of having been the first to prove the intermediate-value theorem for continuous functions. Cauchy's proof owes its mechanics to a particular approximation procedure very simple in conception. Nevertheless, though the mechanics of the proof are simple, the basic conception of the proof is revolutionary. Cauchy transformed the approximation technique into something entirely different: a proof of the existence of a limit.

In the eighteenth century, mathematicians assumed that all polynomials possessed the intermediate-value property. This property could then be used to find the zeros of the polynomial. In particular, if $P(x)$ is the given polynomial and there exist a and b such that $P(a) < 0$ and $P(b) > 0$, then it was assumed that there was some X

between a and b such that $P(X) = 0$. The quantities a and b were called *limits* of the root X.

Once given a and b, X may be approximated to any degree. Here, for instance, is a method expounded in several eighteenth-century algebras.[66] Given $P(a) < 0$ and $P(b) > 0$, consider $P[(a + b)/2]$. If this is zero, we are done. If not, it must be either positive or negative; assume for definiteness that it is negative. Replace a by $(a + b)/2$ and retain b. This gives a new pair of numbers, which are only half as far apart as the original pair, such that $P(a) < 0$ and $P(b) > 0$. By repeating the process of halving the interval between a and b, the root will be approached to any degree of closeness. The difference $b - a$ provides an upper bound on the error made in taking either a or b to be the true root. Maclaurin, in his exposition of this method, said that when the difference $b - a$ was less than 1, a or b could be used as the first approximation for Newton's method.[67]

A slightly different version of this method was given by Lagrange. Suppose that $P(A)$ and $P(B)$ have opposite signs. Then an n of any smallness may be chosen; let x be successively $A, A + n, A + 2n, \ldots, B - 3n, B - 2n, B - n, B$. Eventually, by simply inspecting the successive values of P, there will be found two successive values in this sequence that give opposite sign. Furthermore, as Lagrange pointed out, the process itself gives bounds on the "error"—that is, the difference between the real root and either of the approximations: the error is less than or equal to n.[68] Lagrange presented his method together with a method of finding a number n that is less than the difference between any two of the real roots of P.[69] When n is so chosen, Lagrange's method finds *all* the real roots between the given limits A and B. It is this feature that made Lagrange's method attractive to mathematicians concerned with finding the roots of equations. But it is the possibility of repeating the division of the interval into parts, not specifically mentioned by Lagrange, that made the method appeal to Cauchy.

Cauchy took this approximation procedure and repeated it—as Maclaurin had repeated the halvings—so as to convert it into a proof of the existence of the intermediate value. (For the exact text of Cauchy's proof, consult the appendix.) In Cauchy's notation, let f be a continuous function on some interval including the values x_0 and X. If $f(x_0)$ and $f(X)$ have opposite sign, if $X - x_0 = h$, and if we divide the interval $X - x_0$ into m parts each of length h/m, proceed, just as Lagrange had, to consider the signs of $f(x_0), f(x_0 + h/m), f(x_0 + 2h/m), \ldots,$ $f(X - h/m), f(X)$. Two values of $f(x)$ may be found whose signs are opposite and whose arguments differ by h/m.

But Cauchy was not using this procedure just to find the roots of $f(x) = 0$ between x_0 and X within an error of h/m. Instead, Cauchy repeated Lagrange's approximation procedure, applying it now to the new interval of length h/m.[70] By this repetition, he produced two sequences of values of x, one increasing and one decreasing, such that the terms of one sequence get arbitrarily close to the terms of the other. In Cauchy's notation, the sequence of increasing x values is $x_0, x_1, x_2, \ldots$; the sequence of decreasing x values is $X, X', X'', \ldots$; and the sequence of differences is $X - x_0$, $(1/m)(X - x_0)$, $(1/m^2)(X - x_0)^2$, etc. The two sequences $x_0, x_1, \ldots$ and $X, X', \ldots$ must converge to a common limit, which Cauchy designated by a. But $f(x)$ is continuous between x_0 and X; that is, for Cauchy, "For each value of x between those limits, the numerical [i.e., absolute] value of the difference $f(x + \alpha) - f(x)$ decreases indefinitely with α."[71]

Thus, Cauchy argued, the corresponding sequences $f(x_0), f(x_1), f(x_2), \ldots$ and $f(X), f(X'), f(X''), \ldots$ must converge to the common limit $f(a)$. Finally, since the two sequences are of opposite sign (i.e., one is nonnegative and one is nonpositive), their common limit $f(a)$ must be zero. Thus Cauchy had proved the existence of a solution a to the equation $f(x) = 0$ for x between x_0 and X.

Unlike his predecessors, Cauchy was not trying to approximate a root, but to prove its existence. The proof took Lagrange's approximation technique and stood it on its head. The approximation technique supplied the basic first step in the proof: finding two quantities x_1 and X' that give values of f with opposite sign, and showing that these two quantities differ by some precisely specified given quantity that can be made as small as desired by taking m sufficiently large.

Cauchy's proof has three characteristics found in many of his rigorous proofs in analysis. First, it assumes implicitly a form of the completeness axiom for the real numbers: the existence of the limit of a bounded monotone sequence. Such unexamined assumptions are a major, if unavoidable, logical weakness of the *Cours d'analyse*. Second, it leaves some details as an exercise to the reader: for instance, Cauchy did not explicitly calculate the epsilon and k values needed in the argument that if the limit of a sequence $\{x_k\}$ is a, then if f is a continuous function, the limit of the sequence $\{f(x_k)\}$ is $f(a)$. Cauchy often did not bother to make such computations in their full detail; after all, they were familiar to anyone who had studied algebra. Last, but certainly not least, the proof is outstanding in its notation and clarity. For example, the systematic use of index notation—not invented by Cauchy, but often exploited by him—is a great help in making things clear. And the assumptions of the proof are made clearly, not hidden behind a smokescreen of words.

Did Cauchy learn this technique of proof from eighteenth-century algebra, presumably from Lagrange? The answer probably is yes. First, Cauchy proved the theorem in a note to the *Cours d'analyse* entitled "On the resolution of numerical equations." This was almost exactly the title of each of Lagrange's three works on this approximation method: the paper of 1767; the book of 1798; and the fifth chapter of the "Leçons élémentaires," given at the Ecole Polytechnique. Presumably Cauchy was familiar with the "Leçons élémentaires," because that work is the source of the Lagrange interpolation formula, the subject of note V of the *Cours d'analyse* entitled "Sur la formule de Lagrange rélative à l'interpolation." Cauchy certainly had read Lagrange's *Equations numériques* in either the 1798 or 1808 edition, because he referred to it explicitly in his note on numerical equations, a note that, not so incidentally, makes numerous calculations of the bounds on errors in algebraic approximations.

It even may have been a suggestion by Lagrange that motivated Cauchy to give a proof for this theorem. In the text of the *Cours d'analyse*, Cauchy had been satisfied to give a geometric argument for the intermediate-value property of continuous functions,[72] but referred the reader to note III for a "purely analytic and direct" proof. This also had been Lagrange's procedure in the *Equations numériques*; in the text, Lagrange had noted that this theorem was usually proved "by the theory of curved lines," but added—typically—that an algebraic proof would be preferable. He then gave an argument for the theorem based on breaking the equation into linear factors, each of which obviously has the intermediate-value property, but he explicitly recognized that this did not apply to equations with complex roots. Accordingly, he tried to give a better proof in note I to the book.[73] But Lagrange's second proof is not very convincing. It is based on a theorem about continuous functions that is even stronger than the intermediate-value property: if $f(a) < \phi(a)$ and $f(b) > \phi(b)$, there is an intermediate point at which the two functions are equal.[74] Lagrange's proof of this stronger theorem rests on a mental picture of two quantities approaching and passing each other and refers to the physical picture of two bodies moving along the same line while one overtakes the other.* Bolzano was certainly correct when in 1817 he criticized proofs like this as being based on the idea of motion.[76]

It is not hard to imagine Cauchy's interest in an algebraic proof of the intermediate-value theorem being aroused by Lagrange's call for one; his disappointment upon actually reading Lagrange's proof; and the effect of reading immediately thereafter an approximation technique so well suited to the construction of the intermediate

* Lagrange began by representing the proposed equation in general by $P - Q = 0$, P being the sum of terms with plus sign and $-Q$ the sum of those with minus sign. Then, he said, "From the form of the quantities P and Q . . . it is clear that these quantities increase necessarily as x increases, and that, making x increase by all insensible degrees from p to q, they increase also by insensible degrees; but in such a way that P increases more than Q, since the smaller it was the larger it becomes. Thus there will necessarily be a term between the two values p and q where P will equal Q, just as two moving bodies which are supposed to traverse the same line in the same direction and which, beginning simultaneously from two different points, arrive in the same time at two other points in such a way that the one which was formerly behind is later found ahead of the other, must meet on their paths."[75]

value. We have, then, in Lagrange's *Equations numériques* a plausible algebraic source, both as to motivation and as to technique, of Cauchy's proof.

Ivor Grattan-Guinness's suggestion that Cauchy "borrowed" Bolzano's proof of the intermediate-value theorem cannot, on the other hand, be maintained.[77] First, Cauchy had read Lagrange's book, which explicitly calls for such a proof by algebraic methods; thus Cauchy need not have read Bolzano, as Grattan-Guinness contends, to have got the idea of proving the theorem. (Bolzano's motivation was different; he said that a proof was needed to fill a gap in Gauss's otherwise rigorous 1816 proof of the fundamental theorem of algebra.) Second, the proof technique used by Cauchy does not resemble Bolzano's at all. Bolzano began his proof by trying to show that any sequence possessing what we call the Cauchy criterion has a limit. He used this result to prove that a bounded set of real numbers has a least upper bound. He then used the result concerning least upper bounds to prove the stronger theorem about pairs of continuous functions stated in 1798 by Lagrange; and finally, like Lagrange, he derived the intermediate-value theorem as a corollary of that stronger result. I cannot see any way that Bolzano's proof could be converted into Cauchy's. Nor is the hypothesis of Bolzano's influence even necessary, since Lagrange's work on approximation provides a far more plausible context for Cauchy's proof. Indeed, Cauchy immediately followed his proof with a discussion of how the same technique could be used to estimate the error in the corresponding approximation.

What, then, has been shown about the origin of Cauchy's work? Earlier arguments very similar to those used by Cauchy are encountered again and again. Beyond a doubt, these arguments were known to Cauchy, as were their authors. Since Cauchy's references are usually not very specific, however, often nothing further may be inferred. In the present case, I would not wish to make any further inferences; I do not believe that Cauchy looked at Lagrange's work on approximations, said "Aha!," and proceeded to construct his proof of the intermediate-value theorem. This does not seem to be the way Cauchy worked. He tended to be stimulated by some problem, or by hearing or reading something about a subject, to work out vast numbers of results on his own.[78] He sometimes even rediscovered and republished theorems that he himself had previously proved, which justifies calling him a reinventor rather than a copyist. Even an explicit reference by Cauchy to another mathematician cannot prove direct influence, since he could have derived a result and then have had someone else's work called to his attention; courtesy then would require a reference. Even in the case of explicit references to a closely analogous piece of work, as in the case of Lagrange's *Equations numériques*, Cauchy could have been relying simply on recollection. I believe that all that can be documented from the available sources is the climate of mathematical opinion, which shaped Cauchy's questions, and the techniques available. The use he made of them seems to have been supremely his own. He is no less a great mathematician for having been influenced by his time. His achievements were not obvious consequences of eighteenth-century developments. For instance, Lagrange already had available to him all the materials needed to prove the intermediate-value theorem: the desire for a proof, the technique of finding the intermediate value; the interest in rigor; and the technical facility with inequalities. But he did not bring them all together—nor did any other mathematician of the period. Nobody but Bolzano was able to come close to the rigor of Cauchy's work, and Bolzano did not equal Cauchy in the scope of his mathematical achievements until after becoming familiar with Cauchy's work.

Conclusion

Besides believing that the calculus could be made rigorous if reduced to algebra, Cauchy appreciated how the algebra of inequalities used in approximations could provide a rigorous foundation for real analysis. Ever since Gauchy, results in the algebra of inequalities have been presented explicitly as preparation for the rigorous exposition of the calculus. Cauchy himself began this practice in note II to the *Cours d'analyse*, where he systematized and provided a number of specific inequalities needed for his proofs. This note includes what is now called the Cauchy–Schwarz inequality, stated for sums of squares of real numbers.[79] Some of the specific inequalities already can be found in previous work in algebra;[80] it is their status as a basis for the foundations of analysis that is new.

In particular, in the first rigorous proof about derivatives, Cauchy used one of the inequalities from note II in his *Cours d'analyse*. The technique of the proof had been developed by Lagrange to derive the Lagrange remainder, a result that Lagrange viewed as an error bound in an approximation. Cauchy took Lagrange's technique of proof and justified it by his own definitions; he then was able to use it and the associated inequality to give the first rigorous proofs of theorems about derivatives based on the algebra of inequalitites.

In considering the epsilon-based concepts of limits, continuity, and convergence, it should be remarked that the Greek letter ε used by modern mathematicians—a notation that Cauchy originated and applied in several of his proofs—probably comes from the correspondence between '*epsilon*' and the initial letter of the French word *erreur*. In later work on probability theory, Cauchy in fact used the letter epsilon to stand for *error*.[81] The epsilon in a modern proof may be regarded as an inheritance from the days when inequalities belonged in approximations. The epsilon notation is a reminder that, paradoxically, the development of approximations and estimates of error brought forth many of the techniques necessary for the first exact and rigorous proofs about the concepts of the calculus. Eighteenth-century mathematicians were never more exact than when they were being approximate. And it was algebra that paved the way for the recognition of the link between approximation and rigor.

4 The Origins of the Basic Concepts of Cauchy's Analysis: Limit, Continuity, Convergence

Introduction

Central to Cauchy's successful rigorization of the calculus was his simultaneous realization of two facts. First, that the eighteenth-century limit concept could be understood in terms of inequalities ("given an epsilon, to find an *n* or a delta"). Second, and more important, that once this had been done, all of the calculus could be based on limits, thereby transforming previous results on continuous functions, infinite series, derivatives, and integrals into theorems in his new rigorous analysis. Though there were occasional gaps in his reasoning, he nevertheless far outdistanced his predecessors. And his work provided the necessary groundwork for the eventual complete rigorization of analysis by the school of Weierstrass.

I do not know exactly how or when Cauchy came upon his crucial insight that, by means of the limit concept, the calculus could be reduced to the algebra of inequalities. It seems likely that his appreciation of these facts developed gradually, as he worked with specific integrals, approximations, infinite series, and differential equations. Cauchy began his work in analysis with particular problems. Apparently only when he gave his systematic courses at the Ecole Polytechnique did he first deal with questions of rigor in their full generality.[1] If Cauchy had discovered what I have called the crucial insight all at once, we might have expected some expression of this experience in his writings. After all, Abel, who only received the insight at second hand, expressed its impact upon him in no uncertain terms.[2] We can only speculate on this point; Cauchy has left little evidence behind him. The ever important question of how a mathematician comes to his greatest ideas is often hard to answer on the basis of his published work. All that we can do in the present case is to see how the details and the overall plan of Cauchy's work relate to the work of his predecessors. This will serve to illuminate the necessary mathematical conditions of Cauchy's discoveries; the psychological conditions of his discoveries seem hidden.

There is, however, no mystery about where Cauchy got the idea that the calculus could be based on limits. Many mathematicians, including Newton, d'Alembert, and Silvestre-François Lacroix, had already stated that the limit concept was the basic concept of the calculus. They also had applied their verbally expressed limit concepts to defining derivatives and infinitesimals and to finding the sum of an infinite series. But although the techniques of differentiating, computing ratios of infinitesimals, summing series, and evaluating integrals had already been developed, the validity of these techniques had not been proved by means of the limit concept. Indeed, the outstanding attempt to build a rigorous differential calculus before Cauchy, Lagrange's *Fonctions analytiques*,[3] had not explicitly used the limit concept at all; in fact, Lagrange claimed to have eliminated it from his foundation. And there had been no attempt since Leibniz to define the integral as something other than an antiderivative. A number of eighteenth-century mathematicians had stated the goal of basing the calculus on limits; indeed, they believed they had achieved it. But rigorously founding the calculus on limits is easier said than done. There is a difference between stating definitions that sound right and really understanding the concepts. Even more important, there is a difference between understanding the concepts and actually doing the hard work of proving important theorems using those concepts.

Cauchy was both the first to understand the limit concept and the first to apply it successfully to the calculus. He treated limit, continuity, and convergence first, in his *Cours d'analyse* of 1821; he did not give his theory of the derivative and integral until publishing his *Calcul infinitésimal* in 1823. In examining these works, I shall be seeking answers to the following questions: Could a particular eighteenth-century development have helped to show Cauchy the suitability of the limit concept as a foundation for some part of the calculus? Did this development contain a hint of the appropriate definition of convergence, continuity, derivative, integral? Could this development be transformed by Cauchy into a correct proof of a basic theorem?

The Question of Cauchy's Sources

Cauchy's program of rigorization required him to bring together a large number of diverse elements: the theory of algebraic inequalities; a set of widely scattered and apparently unrelated properties of derivatives, integrals, series, infinitesimals, and continuous functions; and discussions of the foundations of analysis. The techniques of the algebra of inequalities came in large part from the works on approximations, especially Lagrange's systematic *Equations numériques*. Most of the other information Cauchy needed, and used, came from four other works: Lagrange's *Fonctions analytiques* and *Calcul des fonctions*,[4] and the two editions of S. F. Lacroix's *Traité du calcul différentiel et du calcul intégral*. This is not to say that all the ideas in these works originated with their authors; in particular, Lacroix owed much to Euler and d'Alembert. Furthermore, Cauchy also knew the work of Euler and d'Alembert and, equally important, of Joseph Fourier and André-Marie Ampère. Nevertheless, the resemblances between Cauchy's work and the books of Lagrange and Lacroix in matters not only of concepts but of notation, language, and other details suggest these works to have been Cauchy's principal sources.[5]

Silvestre-François Lacroix (1765–1843) succeeded Lagrange as professor of analysis at the Ecole Polytechnique in 1799. His three-volume treatise was intended to be a compendium of all of the calculus.[6] In part, it was intended to help those students who, not living in Paris, could not learn mathematics by consulting the original papers in the journals.[7] Lacroix's procedure in treating a topic was to summarize all the major work on it by the leading mathematicians. His table of contents contains an extensive, item-by-item bibliography, enabling the reader to evaluate Lacroix's fidelity to his sources.

There were many views about the nature of the concepts of the calculus in the eighteenth century. Lacroix's book reflects this fact; he did not present his materials under any wholly consistent point of view. For instance, at different parts of the work he treated dy/dx as a limit, a ratio of infinitesimals, and the coefficient of the first-order term in a Taylor series, claiming that they could all be shown to be equivalent. Lacroix was proud of his eclectic view; he said his exposition proceeded by means of a *rapprochement* of all the existing methods.[8]

A modern reader might term Lacroix's procedure a *confusion* of methods rather than a *rapprochement*. Such an exposition seems to involve an uncritical acceptance of the mutually contradictory. But this criticism, whatever its logical validity, is historically entirely wrong. Lacroix, like most mathematicians of the time, wanted to show how to solve problems; therefore his *Traité* included whatever techniques were applicable to this end. Precisely because Lacroix's book is a mathematical museum of diverse methods and results, presented in full complexity, it could be of service to Cauchy. Lagrange preferred to give no theory of the definite integral rather than give one of whose validity he was unsure; Lacroix, instead, told all he knew about eighteenth-century work on definite integrals. Because of Lacroix's attitude, his books were a gold mine for one who, like Cauchy, could identify the gold—the essential defining properties, the techniques that could be generalized and used in proofs—amidst the chaos.

The Origins of Cauchy's Definition of Limit

Introduction

As we have seen, Cauchy defined the limit concept in these words: "When the successively attributed values of the same variable indefinitely approach a fixed value, so that finally they differ from it by as little as desired, the last is called the *limit* of all the others."[9] This concept, translated into the algebra of inequalities, was exactly what Cauchy needed for his calculus. The very language of this verbal definition is sometimes taken to show the superiority of Cauchy's limit concept over all previous work. Cauchy's definition is free from the idea of motion; it does not depend on geometry; it does not retain the unnecessary restriction, often included in the earlier definitions, that a variable could never surpass its limit.[10] All these features already belonged to the treatment of limit given by Lacroix in 1810, however. Although Lacroix did not explicitly define limits in general, his discussion of specific examples makes clear that his understanding was general. For instance, he defined a to be the limit of the function $ax/(x + a)$ as x increases indefinitely, since the difference between a and the value of that function "becomes smaller as x becomes larger, and *can be made less than any given quantity, however small*, so that *the proposed fraction can approach a as closely as desired.*"[11] And in considering the sums of alternating series, he explicitly pointed out, following Simon l'Huilier, that a quantity could surpass its limit.[12]

Upon reflection, we should not be surprised that a reasonable-sounding definition of limit predated Cauchy's work. Had the limit concept not already been fairly clearly defined and freed of unnecessary restrictions, Cauchy might have rejected it, just as Lagrange had done in 1797. Perhaps Cauchy would have recognized the suitability of the algebra of inequalities as a foundation for the calculus even without the suggestion provided by the limit concept; it is of course impossible to know. But the limit concept as understood in 1810, together with a century of statements that the concepts of the calculus could be understood as limits, helped turn Cauchy's attention in the right direction.

Newton Poses the Problem

Newton's *Principia* is the origin of the most important eighteenth-century discussions of the limit concept.[13] The first section of Newton's great book is devoted to deriving lemmas about the relationship between small arcs and straight lines, which then are applied to the mathematical treatment of physical problems. For example, he needed results like this one: If the arc of a curve *ACB* is given, and if the points *AB* approach each other, "the *ultimate ratio* of the arc, chord, and tangent, any one to any other, is the ratio of equality."[14]

The "ultimate ratio," also translated "last ratio," of two quantities approaching zero—or as he later called it, the "limit"—designated a relatively new concept. And although Newton did not formally define the concept, he did explain at length what he meant by these terms. "There is a limit which the velocity at the end of ... [a] motion may attain, but not exceed. This is the ultimate velocity. And there is the like limit in all quantities and proportions that begin and cease to be. And *since such limits are certain and definite*, to determine the same is a problem strictly geometrical."[15] For Newton, then, the concept of limit was clear; the problem was to find them. In Cauchy's time, everybody could calculate simple limits; the problem was to define the concept and determine whether various limits existed.

Newton wanted to avoid infinitesimals, which he did not consider rigorous. So he took pains to explain that "ultimate ratios" were neither ratios of infinitesimals nor 0/0. In so doing, he made his most influential statements about the limit concept, in words that were to recur throughout the eighteenth century: "For those ultimate ratios with which quantities vanish are not truly the ratios of ultimate quantities, but limits toward which the ratios of quantities decreasing without limit do always converge." From this we see that Newton saw the limit as a definite fixed number. He added that the ultimate ratios were limits to which the ratios of quantities decreasing without limit "approach *nearer than by any given difference*, but *never go beyond*, nor in effect *attain to*, till the quantities are diminished in infinitum."[16] For Newton, the limit is a *bound*, to which the variable can approach arbitrarily closely but never exceed and only "ultimately" reach.

The weaknesses in Newton's definition were pointed out by Berkeley in 1734, and discussed later by d'Alembert, Maclaurin, and Lagrange among others. Indeed the history of the limit concept until 1810 is the gradual solution of the verbal problems implicit in Newton's explanation: the eventual substitution of algebraic language for Newton's kinematic expressions; the broadening of the limit concept to include variables that oscillate about their limits; and—crucially—the abandonment of concern over whether a variable reaches its limit. But keep in mind that none of these developments, however often they have been discussed in histories of the calculus, are as important to nineteenth-century analysis as the algebra of inequalities—whose history does not really belong to that of the limit concept at all.

From Physics to Algebra

Newton's explanation of the calculus had included terms like *velocity* and *approach*. Bishop Berkeley, however, strongly objected to the idea of motion being used in the calculus. His view was not immediately adopted. Colin Maclaurin, in his refutation of Berkeley's criticism, was content to state that there was no problem in conceiving fluxions as motion. But Jean d'Alembert strongly advocated freeing the limit concept from physics and making it algebraic; motion, said d'Alembert, is "a foreign idea, and one which is not necessary in the demonstration,"[17] because the calculus is essentially algebraic. Remarks like these were made also by Euler and Lagrange.[18] However attractive motion may have been as an explanatory analogy, there was wide agreement by the mid-eighteenth century that motion no longer occupied a central place in the foundations of the calculus. Thus the first major problem with the Newtonian limit concept had been dealt with—at least in theory.

Moreover, d'Alembert practiced what he preached. He gave one of the first algebraic arguments about the differential quotient as the limit of the ratio of finite differences. In computing the slope of the tangent to the parabola $y^2 = ax$, d'Alembert found that the slope of the secant was equal to $a/(2y + z)$.[19] D'Alembert said, "As we can take z as small as desired, we can make the ratio $a/(2y + z)$ approach the ratio $a/2y$ as closely as desired."[20] Thus, he concluded, $a/2y$ is the limit of the ratio $a/(2y + z)$ and therefore is equal to the slope of the tangent.

Arguments analogous to this one appear occasionally in the latter part of the eighteenth century, though without the application to geometry. Mathematicians often found it useful to write expressions like $a + h$ for quantities whose limit was a. This mode of expression makes it possible to prove simple results about limits; but without the explicit use of inequalities nothing very hard can be proved.

Simple arguments of this sort were well established by the time of Cauchy. For instance, Lacroix proved that the limit of a product is the product of the limits. Let p be the limit of P; q, of Q. In general, $P = p + \alpha$, $Q = q + \beta$, where α and β vanish together after passing through "every stage of successive diminution." Since $PQ = (p + \alpha)(q + \beta)$, $PQ = pq + p\beta + q\alpha + \alpha\beta$. Thus we see that "the difference $PQ - pq$ may be made as small as we please by assigning appropriate values to α and β."[21]

Lacroix did not apply this technique to constructing arguments about derivatives, even in cases as simple as d'Alembert's treatment of the tangent to the parabola. Nor did he make explicit computations of how small α and β

must be to ensure that $PQ - pq$ be less than any given quantity. Nevertheless, the limit arguments he gave are important because they exemplify translations of a verbal limit concept into algebraic language, however simple. Moreover, the algebraic expressions show—as mere words could not show—that the difference between a variable and its limit could indeed be made less than any given quantity. Cauchy, familiar from algebraic approximations with methods of actually computing the corresponding inequalities, was able to envision rigorous proofs about limits and convergence—and therefore about all the concepts of the calculus.

Broadening the Limit Concept: Can a Variable Surpass Its Limit?

The statement that a variable could not go beyond its limit recurred throughout the eighteenth century. It was firmly enshrined in the definition of limit in the *Encyclopédie*: "Without the magnitude which is approaching ever being able to surpass the magnitude which it approaches." D'Alembert added the example of the circle as the limit of inscribed polygons to emphasize this point. It is sometimes suggested that Cauchy was the first to remove this restriction, allowing the variable to oscillate about its limit.[22] In 1795, however, Simon l'Huilier had made a special definition for limits that are approached alternately from above and below in order to discuss alternating series, even though for all other cases he still required limits to be one sided.[23] Here is strong evidence for the general lack of serious concern in the eighteenth century about the details of the foundations of the calculus: it took until 1795 to abandon the "never surpassing" restriction of the limit concept even for this specific case.

Lacroix, his attention drawn to the example of alternating series by L'Huilier's work,[24] abandoned the "never surpassing" restriction in general in 1810. He justified abandoning the restriction on two grounds: it ruled out many mathematically important limits, and it was never really used in practice. Lacroix was aware that he was making a major break with the past usage of the term *limit*, but he felt his action was legitimate. Elsewhere he had cited Pascal to show that defining a term is really arbitrary, being "nothing but imposing a name on things which are clearly designated, in terms which are perfectly known."[25]

Remarking that some variables, indeed, never surpassed their limits, Lacroix said that nevertheless it was wrong to include this restriction in the general definition of limit: "One would thereby exclude the ratios of vanishing quantities whose existence is incontestable, and which we find often in analysis."[26] Lacroix added yet another justification, which pertains much more closely to the eventual use Cauchy made of the concept of limit: "The application of limits is made by means of principles whose truth rests only on the possibility of proving that a variable quantity can approach its limit as closely as desired."[27] This statement was truer than Lacroix imagined.

The Crucial Problem: Does a Quantity Reach Its Limit?

Bishop Berkeley's most important objection to the old limit concept was this: a quantity could never be said to reach its limit, not even "ultimately." He asked, "Where there are no increments, whether there can be any *ratio* of increments? Whether nothings [sic] can be considered as proportional to real quantities? Or whether to talk of their proportions be not to talk nonsense?"[28]

In what would have been a reasonable answer to Berkeley's query had they had a clearer understanding of the inequality nature of the limit concept, Maclaurin, d'Alembert, and Lacroix all tried to explain the distinction between the ratio and its limit much as Newton had done. For instance, Lacroix said, "We do not consider the ratio of quantities when they are vanishing, nor do we conceive quantities to have a ratio when they cease to exist. The limit of the ratio is not the ratio itself, but a quantity to which it [the ratio] can approach as closely as desired."[29]

Lacroix's statement is not original. It is derived ultimately from Newton's statement, which was the basis for somewhat clearer statements later on. Maclaurin, for instance, wrote, "There is nothing to hinder us from knowing what was the ratio of those increments at any term of the time while they had a real existence, how this ratio varied, and to what limit it approached, while the increments were continually diminished."[30]

D'Alembert, who knew the work of both Newton and Maclaurin, said, "The ratio is not exactly equal to the limit; and when the terms are zero, there is no longer a ratio properly so called, for there is no ratio between two things which do not exist; but the limit of the ratio that these differences had to each other when they still existed is something else; this limit is not less real."[31]

It is in this context that Cauchy's discussion, not greatly different, on the same point should be read: "While the two terms [of the ratio] indefinitely and simultaneously approach the limit zero, the ratio itself can converge to another limit, either positive or negative."[32] But Cauchy did not, like Newton, raise the question whether the ratio and its limit were "ultimately" equal, or, like d'Alembert, whether a secant ever "became" a tangent. Cauchy knew when to stop. His definition of limit stated *only* that the variable and its limit differed by less than any desired quantity—as Lacroix had said.

All the other mathematicians quoted assumed that the ratio of vanishing quantities converged to a limit, and asked *how* it could do so. Cauchy said only that it *could* converge to a limit, not that it necessarily did. For instance, when he defined the derivative as the limit of the ratio of $\Delta y/\Delta x$ as both Δy and Δx "indefinitely and simultaneously approach the limit zero," he said, "This limit, *when it exists*, has a determined value for each particular value of x."[33] Examples had long been known in which the limits of ratios did not exist.[34] Apparently Cauchy was the first to appreciate that recognizing such cases did not invalidate the general definition of the derivative. Cauchy's treatment was intended to support valid proofs, not—as had so often been the case—merely to make beginners feel comfortable with a difficult concept. His definition required no more than was necessary for his purpose.

Conclusion

The words Lacroix used to define and describe limits do not sound very different from those used by Cauchy. But Cauchy's understanding of the limit concept was quite different; on several occasions when a proof required a limit, Cauchy translated his definition into the language of algebraic inequalities. And he proved harder propositions than those about the limit of a product. When the limit of a complicated expression was to be discussed, Cauchy occasionally—often enough to show us his clear understanding—actually worked out the delta or n corresponding to a given epsilon;[35] the superiority of Cauchy's limit concept over that of his predecessors does not lie only in the explicit definition, but in the use he made of the concept in proofs.

Continuity

Introduction

Cauchy—and independently, his contemporary Bolzano—gave the first rigorous definition of continuous function. Cauchy and Bolzano recognized the power of the inequality characterization of continuity for proving theorems and, equally important, the vital role that the property of continuity could play in analysis. With his definition of continuity, Cauchy proved the intermediate-value theorem for continuous functions; then he used both this definition and this theorem in proofs of the mean-value theorems for derivatives and integrals and of the existence of the definite integral of a continuous function.

Cauchy gave two forms of his definition of continuous function: the first is algebraic; the second uses the more intuitive language of infinitesimals. Since Cauchy had defined an infinitely small quantity as a variable whose limit is zero, the two definitions are in fact equivalent:

> The function $f(x)$ will be a continuous function of the variable x between two assigned limits ["limit" here means "bound"] if, for each value of x between those limits, the numerical [absolute] value of the difference $f(x+a) - f(x)$ decreases indefinitely with a. In other words, the function $f(x)$ is continuous with respect to x between the given limits if, between these limits, an infinitely small increment in the variable always produces an infinitely small increment in the function itself.[36]

He immediately followed this definition with a proof of the continuity of $\sin x$, which serves as both an example of the definition and evidence of how clearly Cauchy understood the concept.[37]

Bolzano's definition was given in slightly different and more precise language, but its meaning is the same:

> A function $f(x)$ for all values of x inside of, or outside of, certain bounds, varies according to the law of continuity only insofar as, if x is any such value, the difference $f(x+\omega) - f(x)$ can be made less than any given magnitude, when ω is taken as small as desired.[38]

The nearly simultaneous discovery of the definition of

continuous function by Cauchy and Bolzano suggests to the historian a possible common debt to the views of continuity put forward by the analysts of the eighteenth century. A large number of properties related to the modern concept of continuity were discussed in that period and, at various times, called by the name *continuity*. A function could be called continuous if it had the intermediate-value property; if it was differentiable; if it was representable by a unique formula; if it made "no jumps"; or if, given an "insensible" change in the independent variable, it underwent only an "insensible" change itself. Sometimes these properties—which we now regard as distinct—were viewed as equivalent; sometimes, though sharing a name, they were seen as different.

The achievement of Cauchy and Bolzano was threefold. First, they were able to pick out the essential property—essential in the sense of capable of supporting proofs about other properties—of continuous function from earlier characterizations. Second, once this property had been chosen, they went beyond earlier language to give continuity a precise, inequality-based definition. Finally, they used this defining property as the basis for proofs of theorems about continuous functions.

Eighteenth-Century Views of Continuity for Simple Functions

The earliest eighteenth-century discussions of properties of continuous functions were confined to well-behaved functions. The concept of continuity was appealed to not to distinguish between continuous and discontinuous functions but only when some "obvious" fact about continuous functions was needed. For instance, the property of being continuous was used in geometry to show that a curve that had a point on each side of some line must intersect with that line.[39] In algebra, polynomials were taken to be continuous functions, so that finding bounds on the root of a polynomial could be taken to imply the existence of that root. It was the recognition that this was an assumption that eventually led Lagrange to try to prove the intermediate-value property.[40] In the course of his proof, Lagrange needed to give a description of continuity for a polynomial with positive terms, and he did it as follows. If P and Q are polynomials with positive terms, defined between $x = p$ and $x = q$, he said, "It is evident that these quantities increase necessarily according to the way x increases, and, when x is increased by all insensible degrees from p up to q, the former [P and Q] increase also by insensible degrees."[41] This, though limited to polynomials, was a first approximation to the property that Bolzano and Cauchy eventually were to use to define continuity.

The terms *continuity* and *law of continuity* were also used in an entirely different sense in discussions of limits and derivatives. Continuity was sometimes taken to mean that if every element of a given sequence had some property, so did the limit of the sequence.[42] One argument of this type provided a sort of ontological proof of the existence of derivatives. It is by virtue of the "law of continuity," wrote Lacroix, "that the increments, though evanescent, still preserve the ratio to which they have gradually approached before they vanish."[43] Thus continuity seemed to be related in some way to differentiability.

Until the mid-eighteenth century, there did not seem to be any real problem in defining continuity; it retained a geometric character. But neither geometric intuition nor the example of polynomials was enough to disentangle what we now see as the essential property of continuous function from the rest of the existing descriptions, since smooth curves and polynomials enjoy all the properties we have listed. And eventually, defining continuity became an urgent problem.

Extending the Notion of Function

The need to define continuity became pressing because the function concept had changed. When the study of functions began, in the eighteenth century, functions were identified with their representing formulas—usually infinite series. The examples of functions most commonly treated—polynomials and their quotients, exponential functions, trigonometric functions, logarithms—led mathematicians to expect functions to be relatively well behaved: that is, to have the intermediate-value property and to possess as many derivatives as needed. These expectations were violated near the middle of the century, however, when the need arose to consider functions that presented themselves not as analytic expressions but as solutions to partial differential equations from physics. These functions might not be given by explicit formulas at all. Once the concept of function had been so extended, mathematicians had to make clear which functions, with which properties, were allowable as solutions to differential equations. This could only be done by finding new ways of classifying functions.

The debate over what kinds of functions were allowable in solving differential equations began when d'Alembert stated and solved the differential equation for the vibrating string. The debate eventually involved d'Alembert, Euler, Daniel Bernoulli, and Lagrange—the leading mathematicians of the century.[44] In 1747 d'Alembert gave the differential equation for the motion of the vibrating string, which is now called the wave equation and usually written as

$$\partial^2 y/\partial x^2 = 1/c^2 \partial^2 y/\partial t^2.$$

D'Alembert was able to show that the solution y must have the general form

$$F(x + ct) - F(x - ct).$$[45]

But this form of the solution immediately posed a problem. What, other than the boundary conditions, were to be the restrictions on the functions F that represent the shape of the string? Must they all be, as d'Alembert believed, definable by one and the same formula throughout? Or, as Euler insisted, could the initial shape of the string be any arbitrary shape—since, for instance, the initial position of a plucked string is an inverted V, whose functional representation would lack a derivative at one point at least? D'Alembert said that the vibrating-string problem could never be mathematically treated unless "the different shapes of the vibrating string be included in one and the same equation."[46] Euler, however, cared more about the generality of the solution to the physical problem than about its mathematical amenability. In fact, when Daniel Bernoulli suggested a trigonometric series as the solution to the differential equation of the vibrating string, Euler objected. Euler believed that functions represented by trigonometric series would act too much like trigonometric functions: they would be always differentiable, periodic, and, in the case of sines, odd; thus they would not be general enough to give all the physically possible solutions. Euler did not fully understand the relevant properties of trigonometric series, but the reason for his stricture is nevertheless both defensible and important; he wanted the definition of function to be as general as possible.[47] An idea of the way the debate over the vibrating string generated a need for methods of characterizing the different types of functions can be gained by looking at one attempt to resolve it—the St. Petersburg Academy prize competition of 1787.

An Attempt at Clarification: The St. Petersburg Prize Essay

In 1787, the St. Petersburg Academy proposed as its prize problem the question of which functions could be used in solving partial differential equations. The language of the prize proposal reflects the language of the debate then raging: "Whether the arbitrary functions which are achieved by the integration of equations with three or several variables represent any curves or surfaces whatsoever, be they algebraic, transcendental, mechanical, discontinuous, or produced by a voluntary movement of the hand; or whether these functions include only curves represented by an algebraic or transcendental equation."[48]

The prize was won by L. F. A. Arbogast. He helped clarify the situation—at least temporarily—by introducing a consistent terminology, by means of which he concluded that any function which we would now call piecewise continuous was acceptable as the solution to a differential equation. Arbogast introduced a new term—*discontiguous*—for a function which had "the parts detached from each other." He used *discontinuous* in a way borrowed from Euler[49] to denote a function that, though not (for Euler) necessarily discontiguous, could not be written as the same algebraic formula on all intervals.

What Cauchy and Bolzano later called *continuous* was essentially what Arbogast called *contiguous*; the way Arbogast explained the term is more important than the term he used. He began by citing a remark in the vibrating-string debate made by the Marquis de Condorcet[50], that the functional solution of a differential equation, even if made up of portions of curves expressed by different formulas, had to have pieces that held on to each other. But what did "hold on to each other" really mean? Arbogast gave three characterizations of this property, which he called *contiquity*. First, he said that "the last ordinate of the old form, and the first of the new, are equal to each other, or differ only by an infinitely small quantity."[51] Also, he said, "Assuming that the variable increases continually, the function receives corresponding variations."[52] Furthermore, he said, the ordinate y, where y is a function of x, "cannot pass brusquely from one value to another; there cannot be a jump from one ordinate to another which differs from it by an assignable quantity."[53] All these characterizations were intended to rule out a function with what we would call a jump discontinuity at some point. But in giving these properties, he was calling attention to the essential property of continuous functions at a point. Furthermore, he linked these three versions of the "no-jumps" property with the intermediate-value property, saying that such functions had to obey what he called the "law of continuity"—"A quantity cannot pass from one state to another without passing through all the intermediate states subject to the same law."[54]

Grattan-Guinness has pointed out, in arguing for Cauchy's dependence on Bolzano, that both Cauchy and Bolzano spoke of continuity over intervals, not just at points. But this manner of speaking was common in the eighteenth century. Euler had already established that in his usage, continuity—that is, being representable by one formula—was a property of the function itself, and not of the particular formula chosen for this or that interval. Lagrânge's description of the continuity of a polynomial focused on the interval between two numbers that gave it opposite signs. And Arbogast, to make clearer the distinction between continuity and contiguity, gave as an example the function defined as

$$\begin{array}{ll} \sqrt{mx} & \text{from } x = 0 \text{ to } x = a \\ 1/k\sqrt{n^2 - (x-p)^2} + q & \text{from } x = a \text{ to } x = b \\ \sqrt{k^2 - (x-r)^2} + h & \text{from } x = b \text{ to } x = c. \end{array}$$

Appropriate choice of the constants can make this function contiguous on $[0, c]$; it is already continuous on each of the three separate intervals.[55]

Arbogast's paper of 1787 can be taken to represent the usual level of understanding at that time of the properties of what we now call continuous functions. The paper and the debate which stimulated it certainly would have provided food for thought for mathematicians like Bolzano and Cauchy. Unfortunately, Bolzano and Cauchy cite neither Arbogast nor any of the other contributors to the debate in connection with their definitions of continuous function. So even though Arbogast's memoir won a major academy prize and therefore may be assumed to have been widely known, there is no evidence that it had a direct influence on Bolzano or Cauchy. Still, one result of the vibrating-string debate was that there were many descriptions of continuous functions in the period around 1800 and thus available to Bolzano and Cauchy. S. F. Lacroix, for instance, wrote in 1806 about such functions, "The smaller the increments of the independent variable, the closer the successive values of the function are to each other."[56] Bolzano in 1830 actually cited another statement from the very page on which this passage appeared.[57] Jourdain is thus certainly correct in stating that the vibrating-string debate called attention to the various properties of continuous functions, even though he did not demonstrate the debate's direct influence on Cauchy.

Cauchy's Early Views of Continuity

In the introduction to his 1814 paper on definite integrals, Cauchy observed that the theorem $\int_a^b f(x)\,dx = F(b) - F(a)$, where $f(x) = F'(x)$, is true "only in the case of a function found increasing or decreasing in a continuous manner between the limits in question. If, when the variable increases by insensible degrees, the function found passes suddenly from one value to another, the variable remaining included between the limits of integration, the differences between each of the brusque jumps that the function makes necessitates a correction."[58] (This passage sometimes is cited to show why Cauchy's thinking shifted away from the integral as the inverse of the derivative to the integral as a sum.[59]) The language of his description of discontinuities—"pass brusquely," "jumps"—is very much like Arbogast's and is doubtless written with awareness of the eighteenth-century debate. His use of the phrase "insensible degrees" is reminiscent of Lagrange's *Equations numériques*, whose characterization of the continuity of a polynomial may well have helped suggest the crucial property of continuity.

Cauchy did not employ, at least explicitly, any of the "competing" characterizations; he made no use of the equation, if any, representing a function in his discussion of the function's continuity, nor did he refer here to the function's differentiability. Nevertheless, Cauchy did not actually define a continuous function in this 1814 paper; the property he needed for his proofs was "no jumps," whereas his proof of the intermediate-value theorem in 1821 required his delta-epsilon understanding of continuity.

Cauchy did give in 1814 an algebraic characterization of discontinuity at a point. He considered the case of a function that "passes suddenly from one fixed value to a value sensibly different from the first." If $\phi(z)$ is the function, Z the point, and ξ "a very small quantity," Cauchy described the situation by writing

$$\phi(Z+\xi)-\phi(Z-\xi)=\Delta.$$[60]

Cauchy's characterization of discontinuity is an algebraic translation of the property of having a jump. Though he did not use the language of limits here, it is possible to infer from this equation that Cauchy knew that a function has a jump at the point Z if the limit, as ξ goes to zero, of $\phi(Z+\xi)-\phi(Z-\xi)$ is *not* zero. This algebraic understanding in 1814 of a discontinuity at a point is obviously consistent with Cauchy's definition of continuity in 1821; presumably without reading Bolzano he could have concluded quite easily that the function ϕ is continuous when Δ *does* go to zero with ξ. Since the two characterizations are not identical, however, and since neither inequalities nor limits are mentioned, it is not certain whether Cauchy already had formulated his definition of continuity by 1814.

Toward the Rigorous Formulation: Lagrange's Inequality

The closest an eighteenth-century mathematician came to the Cauchy–Bolzano definition of continuity was not in the debates about continuity at all, but rather in an algebraic discussion—Lagrange's work in approximating functions by Taylor series. In 1797 Lagrange tried to prove a result useful in approximating functions by means of their Taylor series: If $f(x+h) = f(x) + hp(x) + h^2q(x) + \cdots$, there is an h sufficiently small so that

$$|f(x)| > |hp + h^2q + \cdots|.$$ [61]

Lagrange's proof was not valid. Nevertheless, in the course of his proof Lagrange gave a description of a continuous function that is closer to the modern Cauchy–Bolzano definition than anything theretofore. Suppose given a function hP, where P is a function of x and h.[62] If x is fixed, said Lagrange, then as long as P exists and remains finite, hP vanishes with h. If the curve corresponding to this function is considered, with h as abscissa and hP as ordinate, the curve will cut the axis at the origin. And, Lagrange said, "The course of the curve will necessarily be *continuous* from this point; thus it will, little by little, approach the axis before cutting it, and approach it, consequently, within a quantity less than any given quantity."[63] This characterization of continuity appears geometric. But Lagrange rendered it algebraic: "So we can always find an abscissa h corresponding to an ordinate less than any given quantity; and then all smaller values of h correspond also to ordinates less than the given quantity."[64] This is a far cry from "insensible degrees" or "infinitely small changes." But it is not far from this characterization of continuity at $h = 0$ to the Bolzano–Cauchy definitions of continuity in general. Even though Lagrange himself did not take his characterization to be the defining property of continuous function, he had for the first time stated, in terms of inequalities, what Cauchy and Bolzano later recognized as such.

Lagrange, then, twice had highlighted the essential property of continuous functions. Moreover, his characterization of the continuity of polynomials in 1798—"When x is increased by all insensible degrees from p to q, [the corresponding functions] increase also by insensible degrees"—appears in the context of his proof of the intermediate-value theorem, which was, for both Cauchy and Bolzano, a major application of their definitions. I have already documented Cauchy and Bolzano's familiarity with *Equations numériques*. Further, both Cauchy and Bolzano frequently cite Lagrange's *Fonctions analytiques*,[65] in which the inequality characterization of continuity is found.

In fact, the correspondence between Lagrange's view of continuity and the Cauchy–Bolzano view was pointed out in 1830 by Bolzano himself. Bolzano noted that the term continuous function had been used in several different senses: for instance, by Lacroix to mean that the ratio $\Delta F(x)/\Delta x$ was bounded; by Kästner and Fries to mean that the function has the intermediate-value property; by Eytelwein to mean that the function is real and finite.[66] These properties are all important, said Bolzano, and they need to be distinguished from each other. But he added that the term continuous function should be reserved for the defining property. In his own words, "Thus it would be best, holding to the terminology introduced by *Lagrange*, *Cauchy*, and others, to understand by the continuity of a function only the property described in the previous section":[67] that is, the property that the absolute value of the difference $F(x + \Delta x) - F(x)$ "becomes and remains smaller than any given fraction $1/N$ if Δx is taken sufficiently small."[68]

The Cauchy–Bolzano Achievement

Cauchy and Bolzano cut through the details of earlier discussions of solutions to the wave equation, intersecting curves, polynomials, and power series near zero; they recognized and isolated the essential defining property common to all continuous functions. It was precisely the property each needed in their completely different proofs of the intermediate-value theorem for continuous functions. Equally important, it was precisely the property needed by Cauchy in his later works on the differential and integral calculus. This episode again illustrates Cauchy's twin abilities to pick the crucial property of a concept out of a mass of results and to formulate that property correctly.

But Cauchy and Bolzano did much more than their predecessors. In particular, having chosen the crucial defining property of continuous functions, they initiated the rigorous theory of continuity by proving theorems. Once Cauchy had proved the intermediate-value theorem for continuous functions, he used his definition and the intermediate-value theorem in his work on calculus: for instance, to prove the mean-value theorems for derivatives and integrals; and to prove the existence of the definite integral for what he called a continuous function, but which actually was a uniformly continuous function.[69] Later, mathematicians extended Cauchy's theory of continuous functions. Abel correctly treated the continuity of functions defined by power series. Weierstrass and his school distinguished—as Cauchy had not—between pointwise and uniform continuity. All this became possible because Cauchy had isolated the crucial defining property of continuous function, associated with it a valid and fruitful method of proof, and taught both to a generation of mathematicians through his *Cours d'analyse*.[70]

Convergence

Introduction

The concept of limit is the basis for all of Cauchy's calculus. It is used in the definitions of continuity and convergence in the *Cours d'analyse*, and in the definitions of derivative and integral in the *Calcul infinitésimal*. The concept of continuity is an immediate application of the limit concept, and the basic properties of continuous functions were needed by Cauchy in his study of derivatives and integrals. Though the concept of convergence of series is less fundamental to the calculus than the concept of limit, it is important for two reasons. First, it is Cauchy's most detailed application of his inequality-based limit concept; Cauchy's treatment of convergence is the first practical demonstration of his sophisticated understanding of how the limit concept could be translated into complicated inequalities so as to prove theorems. Second, while Cauchy did not base his calculus on power series as Lagrange had, series, and in particular power series, retain an importance exceeded only by derivative and integral in the theory and applications of calculus from at least the time of Euler until the twentieth century. In fact, Cauchy was the first person fully to appreciate that the theory of infinite series was not merely a minor extension of the algebra of polynomials, but could be given a basis in the theory of limits just like the other concepts of analysis.

In the *Cours d'analyse*, Cauchy did much more than define the sum of a convergent series. He stated the Cauchy criterion, proved that it was a necessary condition for convergence, and stated and illustrated with examples—

though he did not prove—that it was also sufficient. For series with positive terms, implicitly taking for granted what is now called the comparison test, he rigorously proved the validity of a number of general convergence tests, including the root test and the ratio test. He proved formulas for the sum and product of series with positive terms. And he showed which of his convergence results remain valid for series with both positive and negative terms. Finally, he used his theory of convergence tests to compute radii of convergence for series that were functions of real or complex variables. Thus his work laid the foundation for the rigorous study of functions, both real and complex, defined by power series.

All this was both original and influential. Nevertheless, some important features of Cauchy's theory of convergence were gleaned from the work of his predecessors. For example, though Cauchy's basic results are stated for general numerical series, his choice of material seems to have been motivated by the example of power series and reflects the traditional preoccupation with this kind of series. This is evident especially from his statements about the importance of the root test[71] and the space he gives to finding radii of convergence.

Eighteenth-century mathematicians had given formal derivations of many results—sums, product, inverses—for power series,[72] and they used power series in solving problems of many types: finding roots of algebraic equations; solving differential equations; and evaluating definite integrals. Most important, Euler and Lagrange had established the fruitfulness of studying functions by means of their power-series expansions. Lagrange even had tried to use power series to provide a foundation for the calculus. The *Fonctions analytiques*, in both its formal manipulations of power series and its unprecedently careful attention to remainders, shows what could be done with power series using existing methods. But, though Lagrange derived the first explicit forms for the remainder term of the Taylor series, he provided no general theory of convergence.

The eighteenth century influenced Cauchy's conclusions on general series in ways other than just providing the motivating example of power series. There was the work on error bounds in algebraic approximations, the anomalous results that sometimes appeared in work with divergent series, and the wealth of results concerning infinite series in all areas of mathematics derived by Newton, Euler, Laplace, and Lagrange. Also, there was the work of Lacroix; there are some direct resemblances between Cauchy's discussion of convergence and the discussion of earlier work on infinite series in Lacroix's three-volume *Traité du calcul.*

Cauchy's debt to his predecessors should not be overestimated, however. The rigorous parts of their work on series had been restricted largely to particular cases. The error bounds and ratio test given by d'Alembert, for instance, applied chiefly to the binomial series. Lagrange's work on remainders was developed for the Taylor series. Even Gauss's memoir of 1813, correctly described by many historians as the first entirely rigorous investigation of convergence,[73] considered only one class of series: the hypergeometric. Cauchy's theory was entirely general.

Cauchy's New Definitions

In 1821 Cauchy gave these now classic definitions of convergence and the sum of a series with terms $u_0, u_1, u_2, \ldots$: "Let $s_n = u_0 + u_1 + u_2 + \cdots + u_{n-1}$ be the sum of the first n terms, n being any integer. If, for increasing values of n, the sum s_n approaches a certain limit s, the series will be called convergent, and the limit in question will be called the sum of the series."[74] In giving this definition, Cauchy departed somewhat from the common usage of the term convergent. Since this older usage has been a source of confusion to modern readers, I shall discuss it in detail.

It is often noted with alarm that eighteenth-century mathematicians state that a series is convergent when the nth term goes to zero and divergent only when the nth term does not go to zero. This of course is "obviously" refuted by the example of the harmonic series $1 + 1/2 + 1/3 + 1/4 + \cdots$, whose nth term goes to zero and yet has no finite sum. But it is in fact true if one realizes that it reflects one way of *defining* the term converge. In eighteenth-century work on series, sometimes a series is said to converge in the way that the hyperbola "converges" to its asymptote, that is, when its nth term goes to zero;[75] at other times the series is said to converge in our sense, that is, when its partial sums approach a limit, which is then called the sum of the series.[76] Thus a series may converge in the first sense without converging in the second sense—Cauchy's (and our) sense.

That the harmonic series diverges had long been known. It was first shown in the fourteenth century by Nicole Oresme and was rediscovered by Jakob Bernoulli in 1689. By the mid-eighteenth century, it was well known; both Euler and Condorcet, for instance, are thoroughly aware that a series whose nth term goes to zero need not have a finite sum.[77] Cauchy had to fix the sense of the word, and he chose well. The crucial fact about an infinite series is not whether the *terms* converge (that is, to zero) but whether the series itself—that is, the *sums* of terms—converges to a limit. For Cauchy the important thing about an infinite series was whether it yielded a well-defined finite number for its sum. In reaction against uncritical manipulations with divergent series, Cauchy insisted that "a divergent series does not have a sum."[78] (Of course, divergent series can be very useful, and may be dealt with by extending the idea of summability or through the theory of asymptotic expansions. Cauchy's stricture thus may be seen as an overreaction, which temporarily called into question the value of work like Euler's. It should be noted, however, that a rigorous theory of divergent series presupposes a clear understanding of convergent series.)

Cauchy's definition of convergence of a series has had a curious effect on the modern reader's attitude toward eighteenth-century work. On the one hand, the modern reader is shocked to see in this work the apparent identification of series convergence with the nth term of the series going to zero. "Don't these people know about the harmonic series?" On the other hand, seeing, for example, d'Alembert use the ratio of the nth to the $(n + 1)$*st* term of a series to test for convergence—that is, whether the terms get smaller—a modern reader may conclude that the ratio test is being used in the modern way. Because Cauchy was consistently conservative in his terminology and did not explicitly criticize older usage, his kindness to his predecessors may have led later historians astray. Moreover,

Cauchy's choice of terminology does not destroy the value of earlier investigations of convergence, for the series most often treated in them are power series. Except at the end points of the interval of convergence, a power series is convergent in Cauchy's sense if and only if it is convergent in the sense of Klügel and d'Alembert: the nth term goes to zero. Cauchy's theory of convergence in fact preserves the value of many earlier proofs that particular series converge—as of course it was intended to do.

Yet it was not in the definition of convergence that Cauchy's originality lay, nor even in his insistence that calculations with divergent series are not to be relied upon, for the discussion of infinite series in Lacroix's *Traité* expresses essentially the same views. Lacroix began his discussion of series by noting that the series for the function $a/(a - x)$, developed in powers of x, is $1 + x/a + x^2/a^2 + \cdots$, which does not give the correct value unless x is ("abstraction made of the sign") less than a. This example moved him to the reflection, based on a similar remark made by d'Alembert, that to be safe in using such a series development "we must discuss with care the convergence[79] of numerical series . . . and we ought to count on these determinations only when we can assign the bounds [*limites*] of the difference that can be found between these [series] and the true value."[80] Cauchy agreed and put it even more strongly when he said that a divergent series did not have a sum.[81]

Lacroix added that for a series development to be of any use, the difference between the series and its true value should "be made less than any given quantity, no matter how small," as more and more terms are taken. This immediately suggests that the sum of an infinite series should be understood as the limit of the partial sums. This definition of sum of a series is implicit in most eighteenth-century work in series and indeed had been stated explicitly by Colin Maclaurin.[82] Thus, the merit of Cauchy's definition of convergence is not apparent from the definition itself or the repetition of Lacroix's warning against divergent series. Here as in the case of the limit concept, getting the right definition is only the beginning.

The Cauchy Criterion and the Completeness of the Real Numbers

Cauchy's name is associated with the study of the completeness of the real number system through the *Cauchy criterion*, which states that every Cauchy sequence converges.[83] Cauchy himself stated the Cauchy criterion for the infinite series $u_0 + u_1 + u_2 + \cdots + u_n + \cdots$. Let $S_n = u_0 + u_1 + u_2 + \cdots + u_{n-1}$. To say that such a series converges according to Cauchy's—and our—definition is to say that "if, for increasing values of n, the sum S_n indefinitely approaches a certain limit S, the series is called *convergent*, and the limit in question is called the *sum* of the series."[84] Cauchy then called attention to the differences between the first and the successive partial sums, defined by

$$\begin{aligned} S_{n+1} - S_n &= u_n, \\ S_{n+2} - S_n &= u_n + u_{n+1}, \\ S_{n+3} - S_n &= u_n + u_{n+1} + u_{n+2}, \\ &\vdots \end{aligned}$$

For the series to converge, it was known to be necessary that the first of these, u_n, go to zero. But it was also known, as Cauchy pointed out next, that this was not sufficient:

> It is necessary also, for increasing values of n, that the different sums $u_n + u_{n+1}$, $u_n + u_{n+1} + u_{n+2} \cdots$, that is, the sums of the quantities $u_n, u_{n+1}, u_{n+2}, \cdots$, taken, from the first, in whatever number we wish, finish by constantly having numerical [that is, absolute] values less than any assignable limit. *Conversely, when these diverse conditions are fulfilled, the convergence of the series is assured.*[85]

Four years before Cauchy published his *Cours d'analyse*, Bernhard Bolzano explicitly stated the "Cauchy criterion," and tried to prove it:[86]

> If a sequence [Reihe] of magnitudes $F_1(x), F_2(x), F_3(x) \cdots F_n(x) \cdots F_{n+r}(x) \cdots$ is subject to the condition that the difference between its nth member [Gliede] $F_n(x)$ and every later member $F_{n+r}(x)$, no matter how far beyond the nth term the latter may be, is less than any given magnitude if n is taken large enough; then, there is one and only one determined magnitude to which the members of the sequence approach closer, and to which they can get as close as desired, if the sequence is continued far enough.

The anticipation of the Cauchy criterion of 1821 by Bolzano in 1817 is one of the most striking examples of the coincidence between the work of Bolzano and Cauchy and one of the strongest points made by Grattan-Guinness in his argument for Cauchy's dependence on Bolzano. Can Cauchy's discovery of the Cauchy criterion be reconstructed without having to assume that he had read Bolzano's work?

There are two parts to the Cauchy criterion: it is necessary for convergence; and it is sufficient. Although nobody but Bolzano seemed to have noticed the sufficiency of the Cauchy criterion before Cauchy, earlier mathematics had already provided several hints about its necessity. Cauchy himself made the necessity look almost trivial. For a series of positive terms, Cauchy said that it was necessary that u_n go to zero, but not sufficient. And, he added, it is also necessary that each of the finite expressions u_n, $u_n + u_{n+1}$, $u_n + u_{n+1} + u_{n+2}, \cdots$ go to zero. Obviously, though Cauchy did not explicitly say so, if these *finite* expressions do not go to zero, then the *infinite* expression $u_n + u_{n+1} + u_{n+2} + \cdots$ certainly cannot; the necessity of the Cauchy criterion may be proved in this way.

What is much more important is the sufficiency of the Cauchy criterion. What may have led Cauchy to state this? Cauchy had now given two necessary conditions for convergence of series with positive terms: the nth term goes to zero; and the Cauchy criterion. Everyone knew that the first condition is not sufficient; it would be natural to ask whether the second is sufficient. Examining some examples certainly suggests that it is sufficient. It also fits with one's intuition of convergence.

Unlike the infinite series of terms $u_n + u_{n+1} + u_{n+2} + \cdots$, the finite expressions $u_n + \cdots + u_{n+k}$ are computable. In a sense, these finite expressions are approximations to the actual value of the remainder of the infinite series. One source of Cauchy's interest in these expressions therefore may have been approximations to the value of the remainder of an infinite series. Following d'Alembert, Lacroix had computed the bounds on such remainders for the particular case of the sum of terms from the $(n + 1)$st on in the binomial series.[87] Lacroix was concerned with *finite*, not infinite, remainders: "Such are the limits above and below of the diverse *approximations* which the series gives."[88] That something like this may have been in Cauchy's mind when he devised the Cauchy criterion is suggested by his first illustration of that criterion. In an argument closely analogous to Lacroix's derivation of the bounding inequalities for the binomial-series remainder, Cauchy showed by actual computation that the convergent geometric progression $1 + x + x^2 + \cdots$ satisfied the Cauchy criterion because the finite expressions $x^n, x^n + x^{n+1}, x^n + x^{n+1} + x^{n+1}, \cdots$ were always included between the bounds x^n and $x^n/1 - x$.[89]

Besides the work of Lacroix, there is another possible source: Euler's proof of the divergence of the harmonic series. Cauchy never mentioned Euler's paper, but a plausible case can be made that Cauchy read it nevertheless. As an example of the necessity of the Cauchy criterion, Cauchy proved the divergence of the harmonic series by showing that the sum of the terms

$$\left(\frac{1}{n+1} + \frac{1}{n+2} + \cdots + \frac{1}{2n-1} + \frac{1}{2n}\right),$$

that is, what we would write as $S_{2n} - S_{n+1}$, is greater than $1/2$ for all n.[90] Cauchy said this was a "new proof" of the divergence of the harmonic series. But Jakob Bernoulli had given essentially this proof more than a hundred years before. Since Cauchy nevertheless treated the divergence of the harmonic series as a known result, he must have been familiar with a proof other than Bernoulli's. In a paper of 1734,[91] Euler had proved the divergence of the harmonic series by showing that what we would write as $S_{nk} - S_n$ is bounded below by a precisely computable finite positive number. Euler's criterion was not stated in general, but merely used for this specific purpose. It is possible that the first proof Cauchy saw of the divergence of the harmonic series was Euler's. Even if this is not the case,[92] he well may have sought out Euler's paper on harmonic progressions for its treatment of a series that converged in the old sense—having diminishing terms—but not in the new. In either case, Cauchy could have formulated the Cauchy criterion by generalizing Euler's test.

However Cauchy was led to the Cauchy criterion, the

fact is that he not only stated it, but used it. He applied the sufficiency of the Cauchy criterion to prove the convergence of the series $1 - 1/2 + 1/3 - 1/4 + \cdots$. Here we are no longer dealing with illustrations: the Cauchy criterion is essential to this proof.[93] Cauchy added—correctly—that his proof could be generalized to show that any alternating series whose terms go to zero is convergent.

In his independent discovery of the Cauchy criterion, Bolzano also may have been influenced by Euler or Lacroix. Since he stated the criterion for sequences, not sums of series, he may have been guided instead by the mental image of points about a cluster point. We cannot know for sure. In any case, Bolzano's attitude toward the Cauchy criterion differed from Cauchy's. Bolzano was interested in properties of the real numbers, like the least-upper-bound property; Cauchy was interested in deriving convergence tests for series. Bolzano needed the Cauchy criterion for his proof of the intermediate-value theorem for continuous functions. Cauchy's own applications—as opposed to illustrations—of the Cauchy criterion are few.[94]

The importance of the Cauchy criterion in the theory of real numbers was not widely appreciated by Cauchy or any of his immediate successors. In the *Cours d'analyse*, Cauchy never explicitly stated any form of the completeness property of the real numbers save the convergence of Cauchy sequences. Implicitly, however, he assumed several theorems about sequences that are equivalent to, or consequences of, the completeness of the real numbers. Cauchy used these in proofs in his *Cours d'analyse* without justifying them and without ever stating them in general:

1. monotone-sequence property—every bounded monotone sequence converges to a limit;
2. comparison test—if a given series with positive terms is, term-by-term, bounded by a second, convergent series, then the given series is also convergent;
3. lim sup—a bounded sequence has a limit superior (called by Cauchy "the greatest of the limits").

Item (1) is used in the proof of the intermediate-value theorem for continuous functions. Item (2) plays an important role in proofs of convergence. Item (3) is used in the statement and proof of the validity of the root test for convergence of series.

Presumably the reason Cauchy did not recognize these properties explicitly and state them in general is that they appear to be "obvious." In addition, (2) was sanctioned by long usage.[95] Cauchy did state the Cauchy criterion explicitly, doubtless because he needed to use the details of that criterion in his proof of the convergence of the alternating harmonic series, and because it did not seem as self-evident in application as the other criteria. But he did not try to prove the Cauchy criterion in general. He may have viewed its proof as "obvious"—or as impossibly difficult. "Obviousness" is of course partly a function of time; the completeness of the real numbers would no longer be obvious to Dedekind or Weierstrass.

Convergence Tests for Series with Positive Terms

Cauchy followed his definition of convergence with a proof of the well-known result that a geometric series whose ratio has absolute value less than one is convergent according to his definition.[96] This theorem was more than just an illustration; the convergence of the geometric progression was needed to prove the validity of other convergence tests by means of the comparison test. D'Alembert had proved the convergence of the binomial series by comparison with a convergent geometric progression. Cauchy knew d'Alembert's work, if not his paper, since Lacroix followed it closely in his *Traité du calcul.*[97] Cauchy's exploitation of the comparison test left d'Alembert and Lacroix and their particular examples far behind, however, for Cauchy gave tests for the convergence of arbitrary series. Because he had proved their validity by means of the comparison test, they now were free from the particular geometric series used for comparison.

The first theorem Cauchy proved, the root test, was especially important to him since it enabled him later to compute radii of convergence of power series. His theorem states, "Find the limit or limits to which, while n increases indefinitely, the expression $(u_n)^{1/n}$ converges. Let k be the greatest of these limits, or, in other words, the limit of the greatest values of the expression under consideration. The series $[u_0 + u_1 + \cdots + u_n + \cdots]$ will be convergent if $k < 1$, and divergent if $k > 1$."[98] Cauchy's proof, though based on the existing technique of comparison with the geometric series, is vastly superior to anything theretofore.[99] Consider Cauchy's treatment of the case $k < 1$. Cauchy chose U such that $k < U < 1$. Then, for all n sufficiently large, $u_n^{1/n} < U$. (The possibility of so choosing U comes from Cauchy's understanding of the concept that k is the "greatest of the limits.") This inequality implies that for n sufficiently large, $u_n < U^n$. But since $U < 1$, the series $1 + U + U^2 + \cdots$ is known to converge; thus "we may conclude a fortiori" the convergence of $u_0 + u_1 + u_2 + \cdots$.[100]

Cauchy's proof of the root test was followed by his statement and proof of the ratio test, again for series with positive terms: "If, for increasing values of n, the ratio u_{n+1}/u_n converges to a fixed limit k, the series will be convergent when $k < 1$, and divergent when $k > 1$."[101] This test, unlike the root test, had already been used by others in special cases.

Cauchy proved the ratio test in his *Calcul infinitésimal* by appealing, as d'Alembert had done for the case of the binomial series, to the comparison test with the geometric series.[102] Cauchy gave a more elegant and novel proof of the ratio test, however in his *Cours d'analyse*, deducing it from the root test by means of this theorem: "If the series $A_1, A_2, A_3 \ldots$ is such that the ratio of two consecutive terms, A_{n+1}/A_n, converges constantly, for increasing values of n, to a fixed limit A, the expression $(A_n)^{1/n}$ converges at the same time to the same limit."[103]

Cauchy concluded his treatment of series with positive terms by proving several other useful theorems. Most important, for series with positive terms he defined what is now called the Cauchy product of two series and proved, with a careful use of bounding inequalities, that it converges to the product of the separate sums. That is, if

$$S = u_0 + u_1 + \cdots$$

and

$$S' = v_0 + v_1 + \cdots,$$

then

$$(u_0v_0) + (u_0v_1 + u_1v_0) + \cdots$$
$$+ (u_0v_n + \cdots + u_{n-1}v_1 + u_nv_0) + \cdots$$

is a new convergent series whose sum is SS'.

This result is clearly motivated by Cauchy's concern with power series. In the eighteenth century, the Cauchy product had often been used in multiplying power series;[104] Cauchy was the first not only to prove it but to specify the conditions for which it holds. And the proof is important for a reason more profound than its novelty. Products of infinite series were used, and the theorem assumed, even for power series that may have negative terms. Cauchy's proof requires that the terms be positive; presumably he noticed this fact when he tried to prove the theorem for series of mixed signs. Eventually he proved the theorem only under the condition that the series of the absolute values of the terms converges—that is, in modern terms, is absolutely convergent. And he gave an example in which the Cauchy product of two series that are convergent, but not absolutely convergent, produce a divergent series—the first such example to be presented.[105]

Series with Terms of Differing Sign

Eighteenth-century mathematicians usually had dealt with the problem of convergence of series with mixed signs by using the phrase "abstraction being made of the sign." Cauchy tried instead to distinguish the cases in which his general results continued to hold from those in which they might not. The generality of Cauchy's approach is novel; astonishingly enough, though facts like the divergence of $\Sigma\, 1/k$ and the convergence of the corresponding series $\Sigma\, (-1)^{k+1}/k$ had long been known, apparently nobody before Cauchy had seen that the problem of convergence for series with both positive and negative terms needed any type of general discussion.

Cauchy began by stating the theorem that a series converges when the series of its absolute values converges.[106] Then he explained the way in which his convergence tests applied—or did not apply—to series with terms of mixed signs. For instance, if $|u_n| = \rho_n$, the root test becomes "Let k be the limit to which, while n increases indefinitely, the greatest values of the expression $(\rho_n)^{1/n}$ converge. The series $[u_0 + u_1 + \cdots + u_n + \cdots]$ converges if

$k < 1$, diverges if $k > 1$."[107] This theorem makes possible the computation of radii of convergence of power series. In addition, the root test provides the best way of determining what we now call the absolute convergence of a series.

Cauchy concluded his chapter on the convergence of real-valued series by discussing how to find the radii of convergence, his choice of results about numerical series appearing to have been strongly motivated by a concern with power series, the type of series most intimately related to eighteenth-century calculus. Using the root and ratio tests systematically, he computed radii of convergence for a number of well-known series. D'Alembert and Lacroix had done this for the binomial series; Cauchy also gave this example, but now it was just one of many.[108]

Conclusion

In the theory of convergence even more than in the theory of continuity, Cauchy broke new ground and set old results on a new, rigorous foundation. Given his clear, inequality-based understanding of the prevailing concepts of limit and sum of series, Cauchy was able create from a few earlier techniques a new, general theory of series. We may thank Lacroix for having made available methods for computing partial sums, using the ratio test, and comparing series with convergent geometric progressions. We may thank all the eighteenth-century practitioners of the analysis of the infinite, especially Euler, d'Alembert, and Lagrange, for having created a wealth of results and techniques for infinite series, especially power series. We may thank Lagrange not only for his work on power series but for his aid in developing the algebra of inequalities in approximation theory. But recognizing the usefulness of the techniques, making the correct generalizations, and above all exploiting the methods to prove general theorems about series—these were Cauchy's own achievements. His convergence tests, in particular, gave rise to an entirely new subject.[109] He helped found the rigorous theory of power series, essential to nineteenth-century analysis. It was above all Cauchy's work on infinite series in the *Cours d'analyse* that so inspired the young Abel and provided the first great contrast between Cauchy's new rigor and the older analysis.

Applications of Cauchy's Theory of Series

Cauchy's major goal in his theory of convergence was to develop the general theory of series, both real and complex, in a rigorous way. As he said in the introduction to the *Cours d'analyse*, "As for methods, I have sought to give them all the rigor required in geometry."[110] This meant, he said, rejecting arguments about divergent series, or automatic extensions of results from the real to the complex case. Once this had been done, "Propositions . . . requiring the happy necessity of putting more precision into theories, and involving useful restrictions of assertions stated too broadly [trop étendues], will profit analysis, and furnish many subjects for investigation which are not without importance."[111] Among the ways Cauchy showed the importance of having a rigorous general theory was by citing occasional counterexamples to assertions stated without "useful restrictions": for instance, showing that the formula now known as the Cauchy product held only for absolutely convergent series, or later on, that there can be more than one function with a specific Taylor series.[112]

The importance of a general theory of infinite series was certainly clear to Cauchy from the work of Euler and Lagrange, who had shown the way in which a great deal of analysis could be derived from series. Had the demonstration of particular results been Cauchy's central concern, he surely would have given them more prominence in his exposition. Nevertheless, the derivation of results, both old and new, was certainly one motivating circumstance for Cauchy's theory. Given the large amount of emphasis on complex series in the *Cours d'analyse*, it appears that results in complex analysis were among his major concerns.

Once Cauchy had established the theory of convergence for real-valued series, he introduced complex numbers and complex-valued functions. It was easy for him to extend his methods to discuss the convergence of series of complex numbers and power series in a complex variable. In an extensive treatment covering a hundred and twenty pages, he defined and characterized the elementary complex functions and obtained their series developments using his theories of continuity and convergence; some of his results are relevant to finding the sums of significant power series.[113]

Cauchy characterized the common continuous complex-valued functions of a real variable by their functional equations. For instance, he solved the functional equation $\varpi(x+y) = \varpi(x)\varpi(y)$ for a continuous complex-valued function ϖ, deriving the solution $A^x(\cos bx + \sqrt{-1}\cdot \sin bx)$.[114] He defined convergence and proved convergence tests for complex-valued series by applying his results in the real case.[115] He then used the functional-equation characterization of complex functions to find the sums of various complex-valued power series. For instance, he showed that $(1+x)^\mu$ is the sum of the binomial series $1 + \mu x + [\mu(\mu-1)/2!]x^2 + \cdots$ in the complex case, where x is the complex number $z(\cos\theta + i\sin\theta)$, by showing that it satisfies the functional equation $\varpi(\mu)\varpi(\mu') = \varpi(\mu\mu')$.[116] This argument requires that ϖ be a continuous function of μ. Both in the complex case and in the simpler real case he had treated earlier by the same method,[117] Cauchy's derivation of the binomial series rests on the continuity of $\varpi(\mu) = 1 + \mu x + [\mu(\mu-1)/2!]x^2 + \cdots$ as a function of μ, and therefore on Cauchy's erroneous theorem that an infinite series of continuous functions is continuous.[118] (Incidentally, Cauchy's derivation of the binomial series for both the real and the complex case and, based on this, the series for e^x, provide a much more plausible motivation for Cauchy's erroneous theorem about the continuity of infinite series than that given by Imre Lakatos and Grattan-Guinness,[119] who saw it as a veiled attack on Fourier series. Indeed, Cauchy explicitly referred to that false theorem in his derivation of the real binomial series.[120]) Given that continuity, Cauchy was able to find the sum of the series

$$1 + x + x^2/2! + x^3/3! + \cdots, \tag{4.1}$$

where x is the complex number $z(\cos\theta + i\sin\theta)$, from the binomial series. He later defined e^x for complex x as the sum of this series.[121]

In fact, Cauchy derived the sum of the series (4.1) in the complex case in two ways. His first proof starts with the binomial series, replaces z by az (he required that $|az|<1$), μ by $1/a$, and lets $a \to 0$.[122] (In the real case, this technically resembles the way Lagrange derived the series for $y = a^x$ by evaluating $y = (1 + a - 1)^x = [(1 + a - 1)^n]^{x/n}$ according to the binomial theorem, and then determining the coefficients of the resulting power series.)[123] As an easy consequence of his formula for the sum of (4.1), Cauchy gave power series for $\sin z$ and $\cos z$ for real z.[124]

Cauchy then gave a second derivation of the sum of the series (4.1), based on the fact that

$$1 + x\sqrt{-1}/1! - x^2/2! - x^3\sqrt{-1}/3! + x^4/4! \cdots = \cos x + \sqrt{-1}\sin x \tag{4.2}$$

for x real, which he proved by showing that the series in (4.2) satisfies the functional equation $\varpi(x)\varpi(y) = \varpi(x+y)$.[125] From this, Cauchy once more drew as an obvious consequence the power-series developments for $\sin x$ and $\cos x$ for x real[126] and the series for the exponential function.[127] Incidentally, it is interesting to note that these derivations of the sum of the exponential series in the complex case appear to be the only place in the *Cours d'analyse* at which Cauchy derived the series for the real-valued functions $\sin x$ and $\cos x$. Throughout his discussion of complex analysis, Cauchy showed the power of his complex methods by deriving important formulas in real analysis by looking at the real and imaginary parts of complex formulas. In fact, the specific real series and the characterization of real functions by the functional equations they satisfy occurring earlier in the *Cours d'analyse* seem to be chosen because either they are needed later on in the complex case or they are simple examples of arguments that Cauchy later extended to the much more difficult complex case.

Cauchy's interest in these complex results was, of course, prophetic. It is of great historical importance to note that the Weierstrassian theory of functions of a complex variable depends upon the rigorous theory of convergent power series.

Looking Ahead: The Differential and Integral Calculus

In the *Cours d'analyse* of 1821 Cauchy had put limit, convergence, and continuity on a rigorous basis for the first time.[128] He isolated the crucial epsilon properties of each concept and, using inequality techniques developed in both algebra and calculus, proved theorems about them.

These concepts were now available as a foundation for his theory of the derivative and the integral, first published in his *Calcul infinitésimal* in 1823. In this work, both derivative and integral would be defined as limits, and their properties rigorously proved.

5 The Origins of Cauchy's Theory of the Derivative

Introduction

The derivative is the most important of the concepts of the calculus that are defined as limits. Cauchy's *Calcul infinitésimal* of 1823 first gave his unprecedentedly clear and powerful treatment of derivatives. Yet his treatment owes much to the past. The origin of Cauchy's theory of the derivative resembles in some ways that of the concepts of limit, continuity, and convergence. Just as he had done with the concepts of limit and sum of a series, Cauchy took the old concept of differential quotient, or derivative, and gave it a new, precise meaning. He defined the derivative, in the style of the eighteenth century, as the limit of the ratio of the quotient of differences, but his definition was based on his new, clear understanding of the limit concept. Recall that Cauchy had defined *limit* as follows: "When the successively attributed values of the same variable indefinitely approach a fixed value, finishing by differing from it by as little as desired, the latter is called the limit of all the others."[1]

To define the derivative in terms of this definition of limit, Cauchy considered the limit of the ratio of the differences $[f(x+i)-f(x)]/i$ on an interval of continuity of $f(x)$. [The continuity was needed so that $f(x+i)-f(x)$ and i could both "indefinitely and simultaneously approach the limit zero," or equivalently both be "infinitely small quantities." Cauchy defined "infinitely small quantity" as a variable whose limit was zero. Thus, though Cauchy never explicitly stated this as a theorem, every differentiable function must be continuous.] Cauchy observed, as had many of his predecessors, that although the numerator and denominator of the ratio $f(x+i)-f(x)/i$ went to zero, "the ratio itself can converge to another limit, either positive or negative." But Cauchy added to the work of his predecessors: "This limit, when it exists, has a definite value for each particular value of x; but it varies with x. . . . [It will be] a new function of the variable x. . . . In order to indicate this dependence, we give the new function the name of derived function [*fonction derivée*, our derivative], and we designate it with the aid of an accent by the notation y' or $f'(x)$."[2] The phrase "this limit, when it exists" exemplifies Cauchy's rigorous attitude. Perhaps the qualification "when the limit exists" was motivated only by the behavior of known functions at isolated points, but his language was sufficiently general to open up the whole question of the existence—or nonexistence—of derivatives. And though the definition is verbal, Cauchy translated it into the algebra of inequalities for use in proofs.

Unlike most of his predecessors, Cauchy did much more than simply define the derivative; he used his definition to prove theorems about derivatives, and thus created the first rigorous theory of derivatives. The crucial theorem of that theory concerns bounds on the difference quotient: If $f(x)$ is continuous between $x - x_0$, $x = X$, and if A is the minimum of $f'(x)$ on that interval while B is the maximum, then $A \leqslant [f(X) - f(x_0)]/(X - x_0) \leqslant B$. [Cauchy expressed $\leqslant$ verbally.] In his proof, Cauchy translated his definition of derivative into the language of delta-epsilon inequalities: "Designate by δ and ε two very small numbers; the first being chosen in such a way that, for numerical values of i less than δ, and for any value of x between x_0 and X, the ratio $f(x + i) - f(x)/i$ always remains greater than $f'(x) - \varepsilon$ and less than $f'(x) + \varepsilon$."[3]

In this delta-epsilon inequality—the first appearance in history, incidentally, of the delta-epsilon notation—Cauchy expressed the crucial property of the derivative in terms any modern mathematician would recognize. The only shortcoming of this translation of Cauchy's verbal definition is that he assumed his δ would work for all x on the given interval, an assumption equivalent to that of the uniform convergence of the differential quotient. Nevertheless, his use of the inequality to translate the definition is a major achievement. Cauchy knew exactly what he meant by "the derivative is the limit of the quotient of infinitesimal differences," and he was really the first person in history to know this.[4]

For the derivative as for the concepts of limit, continuity, and convergence, Cauchy disentangled the crucial δ-ε property from the mass of earlier work and exploited that property in proofs; its origin may be found in Lagrange's work on approximations involving derivatives.

By the end of the eighteenth century, Lagrange had derived two crucial properties of $f'(x)$, (5.1a) and (5.1b).

(5.1a) $f(x+i) = f(x) + if'(x) + iV$, where V goes to zero with i.

"V goes to zero with i" means that given any D, i can be chosen sufficiently small so that V is between $-D$ and $+D$. Equivalently, given any D, i can be found such that

(5.1b) $f(x+i) - f(x)$ lies between $i[f'(x) - D]$ and $i[f'(x) + D]$.[5]

Equation (5.1a) will be called the *Lagrange property of the derivative*, because not only was Lagrange the first to state it, but he was the first to use properties (5.1a) and (5.1b), as they are still used today, to derive many of the known results about functions and their derivatives, including the properties of maxima and minima, tangents, areas, arc lengths, and orders of contact between curves.

For Lagrange, (5.1a) and (5.1b) rested on the Taylor-series expansions of functions. Cauchy saw that Lagrange's properties, which involve (save for explicit notation) deltas and epsilons, could be assimilated to his newly improved limit concept. Thus (5.1a) and (5.1b) would be an immediate consequence of defining the derivative as a limit. Once the derivative had been so defined, all that Lagrange had deduced from (5.1a) and (5.1b) would remain valid. Thus a large, already existing portion of the logical structure of the calculus would rest for the first time on a firm foundation.

Cauchy knew Lagrange's books on the calculus well. In addition, in 1806 André-Marie Ampère used both (5.1a) and the associated Lagrangian proof technique in a paper about derivatives whose influence Cauchy more than once acknowledged. Cauchy recognized the value of this Lagrange–Ampère proof technique and exploited it; because he also justified it, he became the first rigorously to prove theorems about derivatives.

Proving theorems about derivatives is important, but it is only one part of the differential calculus. The other part, so well developed in the eighteenth century, includes applying derivatives to the solution of problems: finding tangents to curves, tangent planes to surfaces, radii of curvature, maxima and minima, and so on. A rigorous theory of derivatives ought to be expected to justify these applications; Cauchy's theory did so. Here too Cauchy's predecessor was Lagrange. Lagrange's work is almost, but not quite, acceptable by modern standards. Cauchy's definitions and rigorous proofs of results such as the mean-value theorem for derivatives made Lagrange's demonstrations and applications legitimate.

Euler's Criterion: Infinite Series and Remainders

The work that probably inspired Lagrange not only to state the Lagrange property of the derivative but also to exploit it at length is Leonhard Euler's *Institutiones calculi differentialis*. In this work Euler gave a criterion for when to use a finite number of terms of a power series, neglecting their remainder—that is, a criterion for the usefulness of power-series approximations. Euler explained his criterion in the following way. Given y a function of x, and ω a change in x,

$$\Delta y = P\omega + Q\omega^2 + R\omega^3 + \cdots.$$

"If the increment ω, which is added to the variable quantity, is very small, the terms $Q\omega^2, R\omega^3 \ldots$, also become very small, until $P\omega$ greatly exceeds the sum of all the rest." This is Euler's criterion. Essentially, $P\omega$ can be taken to stand for the whole series in all those computations "where the greatest accuracy is not needed." Euler added that "in many cases to which the calculus is applied in practice, this kind of consideration is very fruitful."[6]

What Euler had in mind when he mentioned "cases to which the calculus is applied" is best illustrated by an application he made himself. He showed that if x is a relative maximum or minimum of y, then dy/dx is zero there. Suppose $y(x)$ is a relative maximum. Then

$$y(x) > y(x+a) = y(x) + a\,dy/dx + (1/2)a^2\,d^2y/dx^2 + \cdots, \tag{5.2}$$

$$y(x) > y(x-a) = y(x) - a\,dy/dx + (1/2)a^2\,d^2y/dx^2 - \cdots. \tag{5.3}$$

In each series, for a sufficiently small, the term in a will exceed all the rest; this means that the sign of the entire series of terms containing powers of a will have the sign of the term $a\,dy/dx$ in (5.2) and $-a\,dy/dx$ in (5.3). Thus the only way both inequalities (5.2) and (5.3) can be satisfied simultaneously is for dy/dx to be zero.[7] Of course, this argument requires that the function y be uniquely the sum of its Taylor series, a point that Euler took for granted.

A Taylor-series treatment of maxima and minima in terms of the signs of fluxions of various orders had been given in 1742 and justified geometrically by Colin Maclaurin.[8] Maclaurin did not base his result, however, on a more general statement such as what I have called Euler's criterion. More important, Euler's derivation of the properties of maxima and minima was intended to be purely analytic, not geometric; in effect, he had given an algebraic theory of maxima and minima based on an approximation and thus based on the algebra of inequalities. This theory would have been an important innovation even if it had not influenced Lagrange. But it did.[9]

It appealed to Lagrange because it was consistent with his general program of founding the calculus without geometry or intuition, but solely on what he called "the algebraic analysis of finite quantities." And Euler's work also fits in perfectly with the specifics of Lagrange's "algebraic" foundation for the calculus. Since Lagrange wanted to base his calculus on Taylor series, he would have been impressed especially by Euler's use of Euler's criterion in cases like that of extrema. In fact Lagrange seized on Euler's criterion and extended it far beyond Euler's few examples. He even tried to prove it;[10] though Lagrange's proof makes no reference to Euler, the many similarities between *Fonctions analytiques* and *Institutiones calculi differentialis* argue overwhelmingly for Euler's influence.[11] Finally, Lagrange explicitly credited this property of maxima and minima (if not its derivation) to Euler;[12] he made no reference to Maclaurin's treatment of the subject. Thus it was from Euler's criterion that Lagrange was led to what I have called the Lagrange property of the derivative.

Lagrange and the Lagrange Property of the Derivative

Obtaining the Lagrange property of the derivative was one of the first tasks Lagrange undertook in his *Fonctions analytiques*. Lagrange had begun this work on the foundations of the calculus by trying to "prove" that any function $f(x)$ had a power series expansion of the form

$$f(x+i) = f(x) + ip + i^2q + \cdots,$$ [13]

where by *fonction* Lagrange meant any *expression de calcul* into which the variable entered in any way.[14] Given such an expansion, Lagrange followed Euler in stating that there was some i small enough so that any term of the series, "abstraction being made of the sign," would exceed the sum of the remainder of the terms in the series. But Lagrange, unlike Euler, tried to prove this fact.[15]

Lagrange began his proof by treating $f(x + i)$ as the sum of two expressions, one depending on i, the other not:

$$f(x + i) = f(x) + iP,$$

where P is a function of both x and i. Analogously defining Q by $P = p + iQ$, R by $Q = q + iR$, and so on, Lagrange gave Euler's criterion in the following form:

(5.4) For i small enough, $f(x) > iP$, or for some i, $ip > i^2Q, \ldots$.

Lagrange appealed to the continuity of iP, iQ, ... to assert that i could be found sufficiently small for any *particular* one of the inequalities of (5.4) to hold. This for Lagrange proved Euler's criterion, since if "we can always give i a small enough value so any term of the series ... becomes greater than the sum of all the terms that follow," then "any value of i smaller than that one always satisfies the same condition."[16]

Euler himself had viewed his criterion as occasionally useful in justifying applications of the derivative. Lagrange, however, recognized the result (5.4) as fundamental, saying "[This result] is assumed in the differential and the fluxional calculus, and it is because of this that one can say that these calculuses are the most fruitful, especially in their application to problems of geometry and mechanics."[17] This quotation is an extraordinarily important statement, both from Lagrange's point of view and from ours.

For Lagrange (5.4) provided the answer to a major question he had raised in the Berlin prize proposal of 1784:[18] How could the differential and fluxional calculuses, with their somewhat shaky hypotheses, nevertheless yield "so many true results"? I think the reasoning behind Lagrange's statement was something like this: The differential and fluxional calculuses allow i to become "infinitely small" or to "vanish" or to "have zero as its limit." Whatever else these phrases may mean, they seem *at least* to require that i be a very small finite number—in particular, small enough so that $|iQ|$ can be made less than $|p|$. Thus any result of the differential or fluxional calculus that requires *no more than* making $|iQ|$ less than $|p|$ should follow, in Lagrange's view, from the truth of (5.4).

Whatever the quoted statement may have meant to him, Lagrange in fact justified his applications of the calculus by appealing to a Taylor-series form of (5.4). That is, after Lagrange had defined $f'(x)$ as the coefficient of i in the Taylor-series expansion of $f(x+i)$,* he translated the Euler criterion into the statement that if $f(x+i) = f(x) + if'(x) + (i^2/2)f''(x) + \cdots$, then $if'(x)$ exceeds, in absolute value, the remainder of the series $(i^2/2)f''(x) + (i^3/6)f'''(x) + \cdots$. Lagrange said that this fact about the remainder is equivalent to the Lagrange property (5.1a), $f(x+i) = f(x) + if'(x) + iV$, where V is a function of x and i that goes to zero when i does.[19]

We, of course, recognize the Lagrange property as the crucial defining property of the derivative. Lagrange did not, but he did recognize its equivalence with Euler's criterion.[20] Thus, though Lagrange's Taylor-series approach to the calculus was incompatible with a *definition* of $f'(x)$ according to the Lagrange property, it did not prevent him from recognizing the fundamental importance of that property.

Furthermore, Lagrange quickly translated (5.1a) into inequalities, a step essential to Cauchy's rigorous inequality proofs about derivatives. For V to go to zero when i does, said Lagrange, meant that some i could be found so that the corresponding value of V, "abstraction being made of the sign," would be less than any given quantity. And he followed this verbal statement with a beautiful treatment in terms of the algebra of inequalities, deriving (5.1b):

Let D be a given quantity that we can take as small as we please. We can then always give i a value small enough for

* Lagrange first defined $f'(x)$ as the coefficient of i in the Taylor-series expansion for $f(x+i)$; then he defined $f''(x)$ recursively as the coefficient of i in the Taylor-series expansion for $f'(x+i)$, and so on. He followed these definitions with a formal proof that $f^{(k)}(x)/k!$ is the coefficient of i^k in the Taylor expansion of $f(x+i)$.

the value of V to be included between the limits D and $-D$. Thus, since

$$f(x+i) - f(x) = i[f'(x) + V],$$

it follows that the quantity

$$f(x+i) - f(x)$$

will be included between these two quantities:

$$if'(x) \pm D.$$[21]

This is precisely the property Cauchy used in proving theorems about derivatives in his *Calcul infinitésimal* of 1823.

Since Cauchy knew Lagrange's works on the calculus, it seems quite likely that his rigorous definition of the derivative of a function was based on Lagrange's use of (5.1a) and (5.1b). Lagrange's view of the facts expressed in (5.1) was not, however, the same as Cauchy's. For Cauchy (5.1b) is equivalent to the definition of $f'(x)$; for Lagrange (5.1b) is merely *one* property of the derivative, and not the most essential—though it is the most useful in applications.

Of course, there is more to the rigorous theory of the derivative than a mere definition. The mathematical value of Cauchy's definition stems not from its logical correctness alone, but from the proofs of the theorems to which Cauchy applied it. In proofs as in definition, however, Cauchy owed much to Lagrange.

Proving Theorems about Derivatives

Introduction: Cauchy's Theorems and Their Sources

(5.5)

Cauchy was the first mathematician to prove theorems on the basis of a rigorous definition of the derivative. One such theorem, of particular importance for Cauchy's calculus, states that if $f(x)$ is continuous between $x = x_0$ and $x = X$, then

$$\min_{[x_0, X]} f'(x) \leqslant \frac{f(X) - f(x_0)}{X - x_0} \leqslant \max_{[x_0, X]} f'(x).$$

[Cauchy's own statement and proof of (5.5) and (5.6) are given in translation in the appendix; the notation on this page is modernized.] The mean-value theorem for derivatives is an easy corollary of (5.5) and of the intermediate-value theorem for continuous functions. If $f'(x)$ is continuous between $x = x_0$ and $x = x_0 + h$, then there is a θ between 0 and 1 such that

(5.6) $$[f(x_0 + h) - f(x_0)]/h = f'(x_0 + \theta h).$$[22]

Theorem (5.5) and its consequences are essential to Cauchy's rigorous theory of derivatives and the logical structure of Cauchy's calculus.[23]

But Cauchy was not the first to state and try to prove theorem (5.5). The statement of the theorem, which I shall call the mean-value inequality, and its corollary (5.6), the mean-value theorem, is to be found in the work of two of Cauchy's predecessors: Lagrange and André-Marie Ampère.

Lagrange used the Lagrange property of $f'(x)$ to prove a lemma that in modern terms states that a function with a positive (or negative) derivative on an interval is increasing (or decreasing) there. Lagrange then used this lemma to derive the Lagrange remainder of the Taylor series, a result that yields (5.6) as a special case. In 1806, Ampère, following Lagrange's lead, used inequality techniques to "prove" that $f'(x)$ satisfies relation (5.5) in the following slightly different form: If f is always finite, and not always zero, between $x = a$ and $x = k$, and if $f(a) = A$, $f(k) = K$, then there is some value of x between a and k for which $f'(x) \leqslant (K - A)/(k - a)$, and there is another value of x for which $(K - A)/(k - a) \leqslant f'(x)$. In proving this fact, which is essentially equivalent to (5.5), Ampère stated and used a specific identity in the algebra of inequalities. The same identity later appears in Cauchy's *Cours d'analyse*, and is used by Cauchy in his own proof of (5.5) in his *Calcul infinitésimal*. Cauchy knew Lagrange's work; it will be shown that he also knew Ampère's.[24]

Lagrange Brings Inequality Proofs into the Calculus

Lagrange's use of inequalities in the calculus was in the same spirit as his use of inequalities in algebra;[25] they figured not in definitions, but in the determination of the finite approximations needed in applications and problem-solving. For Lagrange, even the remainder for a Taylor series was just a way of finding, in general terms, the error made when an infinite series is replaced by a finite approximation. Once again the exact mathematics of the nineteenth century has as an important forerunner the previous century's approximate mathematics.

To find the Lagrange remainder, Lagrange first stated and proved a lemma saying that a function with a positive derivative on an interval is increasing there. His method of proof is one of his most important positive contributions to the rigorous basis of the calculus. Lagrange stated the lemma in his *Calcul des fonctions* thus: "A function that is zero when the variable is zero will necessarily have, while the variable increases positively, finite values of the same sign as its derived function, or of opposed sign if the variable increases negatively, as long as the values of the derived function keep the same sign and do not become infinite."[26]

This sort of theorem is obvious if one draws a diagram. It is remarkable that Lagrange felt called upon to give a proof at all—but we must remember his strong preference for algebraic methods over geometric intuition. And the proof itself is even more remarkable; it is as close to a delta-epsilon proof as can be found in the calculus prior to Cauchy.

The proof begins with the assertion that for any function f, (5.1a) holds: $f(x+i) = f(x) + i[f'(x) + V]$, where V is a function of x and i such that when i becomes zero, so does V. Lagrange justified (5.1a) by his appeal to the position of $f'(x)$ in the Taylor-series expansion for $f(x)$. For us or for Cauchy, (5.1a) just states the defining property of $f'(x)$.

Lagrange then modified (5.1a) to say that given D, i could be chosen sufficiently small so that (5.1b) holds: that is, given D, i could be chosen sufficiently small so that $f(x+i) - f(x)$ would be included between $i[f'(x) \pm D]$.[27] Lagrange clearly appreciated that what is important is the *absolute value* of the difference between $f(x+i) - f(x)$ and $if'(x)$.

His realization in 1801 of the significance of absolute values was an important step forward in the use of inequalities in the calculus.[28] While Cauchy was more explicit in his treatment of absolute values, Lagrange's *Calcul des fonctions* had already shown how to use absolute values correctly in proofs.

Lagrange's proof of the lemma begins by applying the "sufficiently small" i of (5.1b) to various points in the interval over which f was defined:

$f(x+2i) - f(x+i)$ lies between $i[f'(x+i) \pm D]$;
$f(x+3i) - f(x+2i)$ lies between $i[f'(x+2i) \pm D]$;

$\vdots$

The reader may have already noted that once D is given, Lagrange assumed that the same i would always work for any x in the given interval.

Since $f'(x), f'(x+i), \ldots, f'(x+[n-1]i)$ all have the same sign by the hypothesis of the lemma, $f(x+ni) - f(x)$ must lie between the quantities

$$\{if'(x) + if'(x+i) + \cdots + if'(x+[n-1]i)\} \pm niD.$$

Lagrange expressed this conclusion by saying that the telescoping sum

$$f(x+i) - f(x) + f(x+2i) - f(x+i) + \cdots$$
$$+ f(x+ni) - f(x+[n-1]i)$$

"will have for limit the sum of the limits,"[29] that is,

$$if'(x) + if'(x+i) + \cdots + if'(x+[n-1]1) - niD$$

and

$$if'(x) + if'(x+i) + \cdots + if'(x+[n-1]i) + niD.$$

Lagrange noted that since D is arbitrary, it can be taken as less than the value of $[f'(x) + f'(x+i) + \cdots + f'(x+[n-1]i)]/n$, "abstraction made of the sign." He gave no reason for being able to choose such a D, but he probably had in mind the fact that D could be taken as less than the minimum value of $|f'(x)|$ between $x = 0$ and $x = ni$. If D is so chosen, it will certainly be less in absolute value than $[f'(x) + f'(x+i) + \cdots + f'(x+[n-1]i)]/n$.* The

* Lagrange elsewhere had given arguments based on calculating quantities like D on the basis of maximum or minimum properties of $f'(x)$ (see, for instance, the argument about P immediately following), so this explanation is likely. Alternatively, Lagrange might have considered D to depend on i and n, in which case D could be calculated. If this were his reason, the rest of the proof would be invalid, since i, and therefore n, must be chosen *after* D. Incidentally, Bolzano (*Rein analytischer Beweis*, §V) believed that the latter had been Lagrange's rationale and had criticized Lagrange's proof on these grounds.

existence of a non-zero minimum for $|f'(x)|$ requires, however, not only that $|f'(x)| > 0$ but that it be bounded away from zero.[30] Lagrange's hypothesis thus would have to be strengthened for this choice of D to be possible.

Supposing that D has been so chosen, Lagrange concluded that $f(x + ni) - f(x)$ will lie between 0 and $2i[f'(x) + \cdots + f'(x + [n - 1]i)]$.[31] Then he defined P to be the greatest positive or negative value of the n quantities $f'(x), f'(x + i), \cdots, f'(x + [n - 1]i)$. For such P, then, $f(x + ni) - f(x)$ lies between 0 and $2inP$.

Lagrange explained what this last inclusion meant. Suppose we represent any function of z by $f(x + z) - f(x)$ and let $z = ni$. Thus z has the same sign as i. If i is taken as small as desired, n can become as large as desired. His proof then shows that $f(x + z) - f(x)$ lies between 0 and $2zP$, so that the lemma is proved to his satisfaction.

There are impressive features distinguishing this proof from almost all previous proofs about properties of the derivative. Small positive quantities are treated by means of the algebra of inequalities, and a delta-epsilon calculation is undertaken. Lagrange did not finish his proof by saying "The sum of a finite number of positive infinitesimals is positive," appealing to a geometrical diagram, or building an impressive curtain of words; the proof is algebraic. Furthermore, he supplied a respectable amount of detail. He developed an extremely useful technique for going from a property of $f'(x)$ on the interval $[x, x + i]$ to a property of $f'(x)$ on the larger interval $[x, x + ni]$ by treating $f(x + ni) - f(x)$ as the telescoping sum $f(x + ni) - f(x + [n - 1]i) + \cdots + f(x + i) - f(x)$. This procedure reappears in the work of Cauchy.

His proof has several weaknesses, however. It assumes implicitly that $f'(x)$ is both bounded and bounded away from zero; he seems to have thought it enough that $f'(x)$ be finite and never equal to zero. And there are even more serious objections. First, the proof is based on the Lagrange property of the derivative (5.1a), $f(x + i) = f(x) + if'(x) + iV$, where V vanishes with i. Lagrange could prove this only (and even then incorrectly) by using the full Taylor-series expansion of $f(x + i)$ in powers of i; but this requires that *all* the derivatives be bounded. Cauchy overcame this objection by defining $f'(x)$ so as to satisfy (5.1a).

Second, by assuming that one choice of i would make V small for all values of x in the given interval, Lagrange confused convergence with uniform convergence. Cauchy reproduced this error.

The Lagrange Remainder of the Taylor Series

Lagrange had given his lemma in order to find the "limits" (that is, bounds) of the remainder term of the Taylor series. He derived the remainder from his lemma as follows. Let the maximum of $f'(x)$ on a given interval be $f'(q)$; the minimum, $f'(p)$. Define two auxiliary functions g and h according to the equations

$$g'(i) = f'(x+i) - f'(p),$$
$$h'(i) = f'(q) - f'(x+i).$$

The definitions of g' and h' make $g'(i)$ and $h'(i)$ positive for x on the given interval, so that the lemma can be applied. Going from these derivatives g' and h' to their "primitive functions,"[32] and assuming that $g(0) = h(0) = 0$, he obtained

$$g(i) = f(x+i) - f(x) - if'(p),$$
$$h(i) = if'(q) - f(x+i) + f(x),$$

which by the lemma must be positive as long as f' remains finite. Then

$$f(x+i) - f(x) - if'(p) \geqslant 0 \text{ and}$$
$$f(x) - f(x+i) + if'(q) \geqslant 0.$$

Thus

$$f(x) + if'(p) \leqslant f(x+i) \leqslant f(x) + if'(q), \tag{5.7}$$[33]

which sets "limits"—that is, bounds—on the value of $f(x+i)$. This is Cauchy's theorem (5.5).

Note that application of the lemma to finding (5.7) requires that the lemma hold also for the *weak* inequality $f(x+z) - f(x) \geqslant 0$, since all that can be claimed here is $g'(i) = f'(x+i) - f'(p) \geqslant 0$. (Lagrange could have avoided this difficulty had he considered instead the functions

$$g'(i) = f'(x+i) - f'(p) + \varepsilon \text{ and}$$
$$h'(i) = f'(q) - f'(x+i) + \varepsilon,$$

which would yield

$$-\varepsilon i + f(x) + if'(p) < f(x+i) < f(x) + if'(q) + \varepsilon i.$$

Realizing that ε is arbitrary yields (5.7). Cauchy used precisely this procedure.) Unfortunately, Lagrange did not use special notation to distinguish between strict and weak inequalities, nor did he appear to appreciate the full significance of this distinction.

Lagrange repeated the procedure exhibited in the case of (5.7) to obtain the nth-order Lagrange remainder. In general,

$$\begin{aligned} &f(x) + if'(x) + \cdots + (i^u/u!)f^{(u)}(p) \\ &\leqslant f(x+i) \\ &\leqslant f(x) + if'(x) + \cdots + (i^u/u!)f^{(u)}(q), \end{aligned}$$

where p and q are, respectively, the minimum and maximum points of the uth derivative of f on the interval $[x, x+i]$. Lagrange concluded from this that there is a quantity X in the interval such that

$$f(x+i) = f(x) + if'(x) + \cdots + (i^u/u!)f^{(u)}(X).$$ [34]

This is now called Taylor's series with Lagrange remainder. Lagrange here stated without proof, as something obvious,[35] the intermediate-value theorem for continuous functions, necessary for finding X.

For Lagrange, using the Taylor series in the calculus made the remainder term seem essential. He derived this remainder by applying the inequality methods already exploited by him in algebraic approximations. Although he made errors, he deserves credit for introducing and helping develop a method of proof that eventually was to establish rigor in analysis.

What Lagrange did for the Taylor series, Cauchy was to do for the derivative: estimate its value by a sequence of inequalities that bound it. What was for Lagrange just a stepping-stone to a first-order error estimate in the Taylor series became a defining property in the hands of Cauchy. Earlier, however, and in a different way, it had become a defining property in the hands of Ampère.

Ampère's Proofs about the Derivative

André-Marie Ampère is best known for his work in electricity. Nevertheless, he wrote a paper that was important in the history of the foundations of the calculus.[36] As a matter of fact, most of Ampère's early work was in mathematics, and it was on the basis of his mathematical work that he was made a member of the Institut de France in 1814, six years before his epoch-making work on electricity began. Ampère's 1806 paper on derivatives and Taylor series, entitled "Recherches sur quelques points de la théorie des fonctions derivées . . . ," was one reason for his mathematical eminence.

Ampère's 1806 paper is important for the historian of the calculus for several reasons. First, it includes features that historians have credited to Cauchy. One of Ampère's goals was to free the calculus from not only the earlier concepts of limits, fluxions, and infinitesimals but also Lagrange's infinite Taylor-series foundation. In particular, Ampère's paper gives inequality "proofs" about the basic properties of the derivative of a function. It also contains an inequality definition for the derivative—unfortunately, not a satisfactory one.

Second, Ampère's paper relies heavily on the work of Lagrange.[37] This is not only in matters of notation, the use of the term *fonction derivée*, and the concern with the remainder term of the Taylor series, but also in its refinement of Lagrange's technique for proving that a function with a positive derivative on an interval is increasing there. Ampère's first major order of business in this paper was to prove Cauchy's mean-value inequality (5.5). Like Lagrange, Ampère had no way of proving that $f(x + i) = f(x) + if'(x) + iV$, where V vanishes with i; still, Ampère used this property and the inequalities based on it to prove theorems about $f'(x)$.

Even if nothing else were known about the relationship between the work of Ampère and Lagrange and the later work of Cauchy, the resemblances cited would suggest that Ampère's paper linked Lagrange and Cauchy. But in addition Cauchy knew Ampère personally and once had been his student. Cauchy said in the introduction to his *Cours d'analyse* that he had "profited several times from the observations of M. Ampère, as well as from the methods that he has developed in his lectures on analysis."[38] More than once he acknowledged Ampère's assistance in a general way.[39] And he explicitly referred to Ampère's 1806 paper in his own proof of (5.5).[40] Cauchy knew Lagrange's

Fonctions analytiques;[41] it is worth adding therefore that Lagrange himself had called attention to Ampère's 1806 paper in the second edition of the *Fonctions analytiques* (1813), acknowledging therein the kinship between Ampère's method of proof and the one he had already given in his *Calcul des fonctions* (1801).[42]

Unfortunately, Ampère's paper is confusing and poorly organized. On occasion it has been misread as an attempt to prove that every continuous function is differentiable.[43] This misreading is due partly to Ampère himself, who wrote that a derivative "exists" when he meant that it was finite and nonzero,[44] and partly to historians, among whom the prevailing view is that analysis prior to Cauchy lacked rigor and sophistication. For these reasons, Ampère's paper has been not only misinterpreted but neglected. What in fact does it say?

Lagrange not only had discredited earlier definitions of the derivative but had given one of his own, thereby basing the calculus on Taylor's theorem. In effect Ampère asked himself, can the derivative $f'(x)$ be defined independently of Taylor's series? The definition of $f'(x)$ would have to specify $f'(x)$ uniquely and define it in unexceptionable terms. Ampère found what he thought was a suitable property of $f'(x)$ in Lagrange's work on the Taylor-series remainder; the property in question was (5.7), which Ampère adopted as the defining property of $f'(x)$:

(5.8) The derived function of $f(x)$ is a function of x such that $[f(x+i) - f(x)]/i$ is always included between two of the values that this derived function takes between x and $x+i$ whatever x and i may be.[45]

All the rigorous nineteenth-century definitions of $f'(x)$ define it by the ratio $f(x+i) - f(x)/i$ and the inequalities that this ratio must satisfy; Ampère was thus the first to give such a definition. His definition has some major deficiencies, however. First, it defines $f'(x)$ at the point x in terms of its values on the whole interval; thus $f'(x)$ must exist on an entire interval to be defined at one single point. This is much too restrictive (though not as restrictive as assuming that $f(x)$ has an entire Taylor series). Second, there is no reason to believe that any such $f'(x)$ exists at all. Third, it is not clear that $f'(x)$ is the only function that satisfies defining criterion (5.8), though Ampère did try to prove that $f'(x)$ was unique.[46] Ampére believed that he had shown both the "existence"—that is, nonzero finiteness—and the uniqueness of $f'(x)$ and that his definition therefore was justified. We, however, are less interested in his definition than in his method of proving that the derivative satisfied it, for this method, based on the work of Lagrange, was adopted by Cauchy.

Ampère began his work confronted with a problem in logic. To prove that his definition of $f'(x)$ made sense, he first had to prove some facts about $f'(x)$. To do this, he had to characterize $f'(x)$ in a way other than by his defining property. Ampère *introduced* $f'(x)$ as the value of the ratio $f(x+i) - f(x)/i$ "when $i = 0$."[47] The properties he actually *used* in his proofs were the properties Lagrange had used: that is, the inequalities satisfied by the ratio $f(x+i) - f(x)/i$ for arbitrarily small i. In particular, he assumed that $f'(x)$ has what I call the Lagrange property: $f(x+i) - f(x)/i = f'(x) + iI$, where I vanishes with i.[48] Once he had proved that the function $f'(x)$ so characterized had the property expressed by (5.8), he turned around and used (5.8) to define $f'(x)$.

Ampère proved that a function with the Lagrange property also satisfies (5.8) as follows. Let $f(x)$ be defined on an interval from $x = a$ to $x = k$. Let $f(a) = A, f(k) = K$, $a \neq k$, $A \neq K$, and let $f(x)$ be finite.[49] (These conditions mean that the function is well behaved, the interval is not a point, and the function is not a constant.) Using a proof technique like Lagrange's, Ampère undertook to show that there was some value of x on the interval such that $f'(x) \leqslant (K-A)/(k-a)$ and some other value of x on the interval such that $(K-A)/(k-a) \leqslant f'(x)$. This gives (5.8). (This result also implies, among other things, that the derivative cannot be zero or infinite on the whole interval and thus, in Ampère's language, "exists.")

Ampère's proof of (5.8) required an algebraic lemma about inequalities that is almost identical with the inequality result used by Cauchy in his *Cours d'analyse* to prove (5.5):[50]

(5.9) In a given interval $[a,k]$, define $b, c, d, e \ldots$ such that $a < b < c \cdots < e < h < k$, and define $B, C, \ldots H$ such that

$f(b) = B, f(c) = C, \ldots f(h) = H$. Now consider the fractions $B - A/b - a, C - B/c - b, \ldots K - H/k - h$. Among these fractions, we can always find a pair such that one of the pair will be greater than $K - A/k - a$, while the other will be less.

This is not a quotation; Ampère intertwined the statement of this lemma with his proof of (5.8), so that it is impossible to disentangle.[51] Let me indicate how Ampère proved the lemma. If, for instance, $a < e < k$, then $K - A/k - a$ lies between $K - E/k - e$ and $E - A/e - a$. For

$$(K - E)/(k - e) - (K - A)/(k - a)$$
$$= (aE - Ae + Ak - aK + eK - Ek)/(k - e)(k - a)$$

and

$$(K - A)/(k - a) - (E - A)/(e - a)$$
$$= (aE - Ae + Ak - aK + eK - Ek)/(k - a)(e - a),$$

giving two fractions with the same numerator and positive denominators, so that both fractions must have the same sign. If both are positive together, $K - E/k - e > K - A/k - a > E - A/e - a$. If both are negative together, $K - E/k - e < K - A/k - a < E - A/e - a$. This proof may be formalized by induction on the number of fractions; Ampère examined several cases and immediately concluded the general result.

Once he had proved this lemma, Ampère specified that $b - a = c - b = \cdots = k - h = i$; thus the lemma shows that there is some x on the interval $[a, k]$ such that $f(x + i) - f(x)/i$ is less than $K - A/k - a$ and another x on the interval such that $f(x + i) - f(x)/i$ is greater than $K - A/k - a$. Ampère then appealed to the Lagrange property of the derivative to go from this result about finite differences to the corresponding result about $f'(x)$: "Since $f(x + i) - f(x)/i$ becomes equal to $f'(x)$ when $i = 0$, it can be represented in general by $f'(x) + I$, where I is a function of x and i which vanishes with i, and which, therefore, can become as small as desired by taking i sufficiently small."[52] Thus by taking i sufficiently small, Ampère concluded that he could find some x on $[a, k]$ such that

$$f'(x) \leqslant K - A/k - a$$

and some other x on the interval such that

$$f'(x) \geqslant K - A/k - a.$$ [53]

This completes Ampère's proof of (5.8).[54]

The key steps in the proof are the perfectly valid lemma on fractions and the passage from the inequalities for the ratio of finite differences $f(x+i) - f(x)/i$ to the inequalities for the derivative $f'(x)$. The second of these steps is valid *if* the convergence of $[f(x+i) - f(x)]/i$ to its limit is uniform. But Ampère had no more reason to assume that he could find a value of i "sufficiently small" to work for all x in the interval than had Lagrange earlier or Cauchy later.

Ampère's paper is important because it transmitted Lagrange's methods of proof, together with a new and useful lemma on fractions, to Cauchy. Also, it is striking that it advocated defining the derivative $f'(x)$ uniquely and unexceptionably: *not* by the imprecise concepts rejected by Lagrange, *not* in terms of the Taylor series, but by an inequality that $f'(x)$, and $f'(x)$ alone, could satisfy. The choice of the inequalities appropriate to supporting the whole logical structure of the differential calculus was not made by Ampère, however, but by Cauchy.

Ampère in effect used the Lagrange property of the derivative as a basis for a theory of derivatives "freed not only from the consideration of infinitesimals, but also from that of the formula of Taylor."[55] For, since Ampère had shown that the derivative $f'(x)$ satisfies his defining inequality (5.8) *if* it has the Lagrange property, he then—incorrectly—felt himself justified in using the Lagrange property to deduce further theorems—including Taylor's. In effect, Ampère assumed the equivalence of his definition with the Lagrange property. To use the Ampère–Lagrange methods of proof, which rest on the Lagrange property of the derivative, Cauchy would have to define $f'(x)$ so as to justify that property. Cauchy's definition was designed to do just that.

Cauchy's Theory of Derivatives: Proving Basic Theorems

Cauchy's rigorous proofs about $f'(x)$ are vastly more than just the culmination of Lagrange and Ampère's work. Still, viewing them in this light helps explain their form. He defined the derivative so as to have the Lagrange property.

He applied this property—now for the first time justified by a definition—to prove the mean-value inequality (5.5) by means of the Lagrange–Ampère method of proof. Indeed, Cauchy's methods of proof are designed, whether consciously or not, to support the proofs worked out by Lagrange and Ampère.

Cauchy stated the basic theorem (5.5) as follows (Cauchy's proof is given in translation in the appendix): "If, $f(x)$ being continuous between the limits $x = x_0$, $x = X$, we designate by A the smallest, and by B the largest, value that the derived function $f'(x)$ receives in the interval, the ratio of the finite differences $f(X) - f(x_0)/X - x_0$ will necessarily be included between A and B."[56] Like Lagrange and Ampère, Cauchy used (5.1b), which is a translation into inequalities of the Lagrange property of the derivative, to prove this theorem. But for Cauchy the procedure was justified by his own definition of $f'(x)$ as a limit. Cauchy's proof is technically like Ampère's, but much easier to follow. Here is a shortened version.

Given $\varepsilon > 0$, we can find δ such that

$$f'(x) - \varepsilon < [f(x+i) - f(x)]/i < f'(x) + \varepsilon, \text{ if } |i| < \delta. \tag{5.10}$$

Statement (5.10) is valid since it simply is a translation of Cauchy's definition of the derivative into an algebraic inequality—a justification immeasurably superior to those given by Lagrange and Ampère.

Note that Cauchy took his definition of $f'(x)$ for a particular x and applied it to the whole interval; he assumed that given an ε, he could find a δ that works for *every* x in the interval. This assumes that $f'(x)$ is the uniform limit of the quotients $f(x+i) - f(x)/i$ in the interval, a confusion also found in the work of Ampère and Lagrange. The confusion arises from failing to specify the variables on which δ depends.

Cauchy then interposed $n - 1$ new values of the variable x, namely, $x_1, x_2, \ldots, x_{n-1}$, between x_0 and X, in such a way that $(x_1 - x_0), (x_2 - x_1), \ldots, (X - x_{n-1})$ are all less than δ. (Here Cauchy differs from Lagrange and Ampère, whose subintervals are equal.) By applying (5.10) to each subinterval Cauchy obtained

$$
(5.11)\quad
\begin{aligned}
&f'(x_0) - \varepsilon < f(x_1) - f(x_0)/x_1 - x_0 < f'(x_0) + \varepsilon,\\
&f'(x_1) - \varepsilon < f(x_2) - f(x_1)/x_2 - x_1 < f'(x_1) + \varepsilon,\\
&\vdots\\
&f'(x_{n-1}) - \varepsilon < f(X) - f(x_{n-1})/X - x_{n-1} < f'(x_{n-1}) + \varepsilon.
\end{aligned}
$$

If A and B are the minimum and maximum values of $f'(x)$ on the given interval, then each of the fractions in (5.11) is greater than $A - \varepsilon$ and less than $B + \varepsilon$.

Now Cauchy applied his version of Ampère's lemma on fractions (5.9)[57] to the fractions $f(x_1) - f(x_0)/x_1 - x_0$, $\ldots, f(X) - f(x_{n-1})/X - x_{n-1}$, all of which have positive denominators. By combining this result with the telescoping property of the sum $f(X) - f(x_{n-1}) + \cdots + f(x_1) - f(x_0)$ used by Lagrange, he passed from (5.11) to the inequality

$$A - \varepsilon < f(X) - f(x_0)/X - x_0 < B + \varepsilon.$$

But since this is true no matter how small ε is, he concluded that

$$A \leqslant f(X) - f(x_0)/X - x_0 \leqslant B.$$

This completes Cauchy's proof of (5.5).

There are many differences between Cauchy's proof and Ampère's. There is of course the difference that Cauchy defined $f'(x)$ to justify the proof procedure. There are also notational differences: using the delta instead of saying "a value of i" makes it much easier to follow the proof, as does the index notation for the values of the variable. Much more important are the conceptual differences. Cauchy made his hypotheses explicit. His proof is crystal clear. And he understood the difference between $\leqslant$, $<$, and *bounded away from*, as is shown in the last lines of the proof, where he skillfully used epsilons to indicate that certain functions were bounded away from their limiting values. Contrasting Cauchy's proof of this theorem with Lagrange's and Ampère's reveals once again Cauchy's ability to cull precise concepts from ill-defined and hazy work and thereby transform them into models of clarity.

One consequence of Cauchy's theorem (5.5) is a corollary (5.6), the mean-value theorem for derivatives. Cauchy derived (5.6) in the form

$$[f(x+h) - f(x)]/h = f'(x+\theta h),\ 0 \leqslant \theta \leqslant 1,$$

using the intermediate-value theorem for continuous functions, which he had proved in the *Cours d'analyse*. This result (5.6) is of importance to us for not only its rigorous proof but its applications. Cauchy found the mean-value theorem and its higher-order analogue, Taylor's theorem with Lagrange remainder, very useful in applying the derivative to solve problems. Reviewing these applications indicates how Cauchy used his theory of the derivative to make rigorous many major results.

Cauchy's Theory of Derivatives: The Derivative at Work

The differential calculus consists of not only theorems about functions and their derivatives but also applications of the derivative to problems of maxima and minima, the geometry of curves and surfaces, and physical phenomena. Such applications may not seem conceptually as interesting to us as the analytic theory of derivatives, but in the eighteenth century these applications were the major raison d'être for the calculus. To set the differential calculus on a firm basis, Cauchy not only had to prove the basic theorems but justify rigorously the wealth of known results.

Before the work of Lagrange, applications of the derivative usually are justified by the use of diagrams and analogies.[58] For instance, the tangent is the limit (in eighteenth-century terms) of the secants, just as the differential quotient is the limit of the quotient of differences. Cauchy's new definition of the derivative makes possible, at least in theory, the use of the algebra of inequalities and of theorems based on inequalities in all proofs about the applications of the derivative. Yet Cauchy, seeking to "reconcile the rigor of proofs with simplicity of methods,"[59] often used the language of his contemporaries, so that his arguments sometimes at first glance appear less rigorous than they really are. For instance, in computing the tangent to a curve represented by y, a function of x, Cauchy said, "Let us conceive now that the point $(x + \Delta x, y + \Delta y)$ comes to approach indefinitely the point (x,y). The secant which joins the two points will tend more and more to coincide with a certain line which is called the tangent to the given curve, and which touches the curve at the point (x,y)."[60] The slope of the tangent,

Cauchy added, is the limit of $\Delta y/\Delta x$ "when the differences Δx, Δy, become infinitely small."[61]

But familiarity with Cauchy's work leads to the expectation that his understanding of the phrases "comes to approach indefinitely," "tend more and more to coincide," "become infinitely small," was clearer than the statement just quoted suggests. And indeed Cauchy used the details of his theory of derivatives to justify many of the applications of the derivative. Here as elsewhere, Cauchy's choice of words should not obscure the great difference in rigor between his work and the work of the Bernoullis, Euler, d'Alembert, and Laplace.

But Cauchy's work was quite similar to that of Lagrange. Cauchy, like Lagrange, obtained most of his applications of the derivative from properties of $f'(x)$ such as the mean-value theorem

$$f(x+h) - f(x)/h = f'(x+\theta h),\ 0 \leqslant \theta \leqslant 1, \tag{5.6}$$

and its higher-order analogue, Taylor's series with Lagrange remainder,

$$f(x+h) = f(x) + hf'(x) + \cdots + (h^n/n!)f^{(n)}(x+\theta h). \tag{5.12}$$ [62]

When expressions like (5.6) and (5.12) appear in Cauchy's work, the arguments used are often descendants of arguments given by Lagrange.[63] With the mean-value theorem and the Lagrange remainder it is possible to justify most of the common applications of the calculus to problems of geometry and extrema.[64]

For instance, Lagrange followed Euler in using inequalities to find necessary and sufficient conditions for relative minima and maxima. But instead of using the infinite Taylor series as Euler had done, Lagrange used finite Taylor series with Lagrange remainders.[65] In his treatment of extrema, Cauchy first proved from Taylor's theorem with Lagrange remainder that if $f'(x_0) = f''(x_0) = \cdots = f^{(n-1)}(x_0) = 0$ and i is "infinitely small," then $f(x_0+i) - f(x_0)$ has the same sign as $i^n f^{(n)}(x_0)$.[66] He then applied this fact to state sufficient conditions for a function with zero derivative(s) at a point to be a minimum or maximum at that point.[67]

Lagrange also used his remainder to show that the slope of the line tangent to the curve $y = f(x)$ is $f'(x)$.[68]

Since he eschewed the limit concept, he defined the tangent to a curve as a line having one point in common with the curve and such that no other straight line can pass between that curve and its tangent line.[69] With this definition he could treat the tangent as a special case of his general theory of orders of contact between curves—a theory in which he relied heavily on the Taylor series with Lagrange remainder.[70] Since Cauchy defined the tangent as the limit of secants, he did not need to use the mean-value theorem (5.6) or the Lagrange remainder (5.12) to show that the tangent to $y = f(x)$ has slope $f'(x)$. Still, Cauchy did need the mean-value theorem to show that the tangent (if it exists) is parallel to the x axis at a minimum or maximum point of a curve.[71]

In the works of Lagrange and Cauchy are found similar treatments of centers of curvature,[72] involutes and evolutes,[73] and Taylor series and remainders for functions of several variables[74] and their application to the theory of curved surfaces.[75] I cannot demonstrate that Cauchy *consciously* borrowed Lagrange's methods; I can only show a close resemblance. But the resemblance is so striking, and the context and notation so similar, that it is hard to avoid concluding that Lagrange's applications of the derivative are the source for many of Cauchy's.

I do not mean to imply that Cauchy did not go beyond Lagrange in applying the calculus. Cauchy appears to have had a deeper geometric insight into the meaning of many ideas, and he found and justified many new results. For instance, Cauchy did not follow Lagrange in giving a wholly analytic treatment of orders of contact. He defined the order of contact between two plane curves[76] geometrically, by his theory of infinitesimals.[77] He used his definition, theorems deduced from it, and the Lagrange remainder of the Taylor series to prove what Lagrange had made his definition: If n is the first integer greater than or equal to the order of contact a, then two curves F and f with order of contact a satisfy $F(x) = f(x)$, $F'(x) = f'(x), \ldots,$ $F^{(n)}(x) = f^{(n)}(x)$, but $F^{(n+1)}(x) \neq f^{(n+1)}(x)$.[78]

Among the new results Cauchy found and applied is what is now called the generalized (or Cauchy) mean-value theorem: If $f(x)$ and $F(x)$ are continuous and have continuous first derivatives between $x = x_0$ and $x = X$ and $F(x)$ is either increasing or decreasing throughout the interval, then for some θ between 0 and 1

(5.13)
$$\frac{f(X) - f(x_0)}{F(X) - F(x_0)} = \frac{f'(x_0 + \theta(X - x_0))}{F'(x_0 + \theta(X - x_0))}.$$ [79]

He proved this by a technique analogous to the one based on Ampère's work used to derive the mean-value inequality (5.6). Cauchy applied theorem (5.13) to deduce several known results, including L'Hôpital's rule and Cavalieri's theorem.[80]

Nevertheless, it must be concluded that Cauchy owed a good part—though not all—of the mechanics of his applications of the derivative to geometry to Lagrange's *Fonctions analytiques*. There is a major difference, however, between their works on applications of the derivative. The difference is not one of technique; it is one of context and, above all, justification. To be sure, Cauchy was not single-mindedly consistent about giving all his arguments a precise form; sometimes he used intuitive descriptions of limits or infinitesimals. But it is clear that he knew how to justify all his applications and could have done so explicitly had he wished.

Conclusion: Cauchy, Lagrange, and the Rigorization of the Differential Calculus

Cauchy received many suggestions from the work of his predecessors on rigorizing the calculus. In the case of the theory and applications of the derivative more than in any other subject, he found much of the work done for him. And it had been done chiefly by Lagrange. Starting with what I have been calling the Lagrange property of the derivative, Lagrange had been the first to apply the powerful tool of delta-epsilon techniques to the calculus.[81] Cauchy understood that what was for Lagrange and Ampère just a useful method of proof was the essential defining property of the derivative. By turning the Lagrange property into a definition, basing his work on a logically consistent set of definitions, and using his own theories of limit and continuity, Cauchy was able to adapt Lagrange's techniques into his own rigorous proofs. Thus he achieved exactly what Lagrange had said should be done in the subtitle of the 1797 edition of his *Fonctions analytiques*: the establishment of "the principles of the differential calculus, free of any consideration of infinitely small or vanishing quantities, of limits or of fluxions, and reduced to the algebraic analysis of finite quantities."[82]

6 The Origins of Cauchy's Theory of the Definite Integral

Introduction

Cauchy's theory of the definite integral is the last of the major concepts of the calculus whose origins I shall trace. It is a suitable topic with which to end this book for several reasons. It is the last basic concept that Cauchy defined. Also, its understanding requires a clear grasp of the concepts already discussed: Cauchy defined the integral as the limit of sums; key to his theory is the proof of the existence of the definite integral of a continuous function;* and he proved (in what we now call the fundamental theorem of calculus) that the integral was the inverse of the derivative.

Not surprisingly, some important techniques in Cauchy's work on integration came from not the eighteenth-century theory of the integral—there really had been no such thing—but from approximations to the values of definite integrals. Nevertheless, Cauchy found comparatively little of the work already done. His predecessors generally had confined themselves to working out foundations for the *differential* calculus, in the expectation that the integral—since it was the inverse of the derivative—would take care of itself.

The Integral before Cauchy: General Views

The integral was viewed quite differently in the eighteenth century than it is today. It had no independent definition of its own; instead, integration was defined as the inverse of differentiation.[1] Accordingly, the indefinite integral was considered to be more fundamental than the definite integral. The definite integral of a function $f(x)$ evaluated between $x = a$ and $x = b$ was by definition $F(b) - F(a)$, where F is defined by the differential equation $dF/dx = f(x)$.

To be sure, Leibniz had defined the integral as a sum; he even had chosen the integral sign $\int$ to be a stylized S for

* Actually, his proof implicitly assumed the function to be uniformly continuous, though he did not distinguish between continuity and uniform continuity, just as he had not distinguished between convergence and uniform convergence.

sum.[2] But most mathematicians rejected Leibniz's definition of the integral as an infinite sum of infinitesimals. Such a definition involves both infinites and infinitesimals, which are problematic concepts at best.[3] Instead, Euler, the Bernoullis, Lagrange, and Laplace preferred to think of integration as the inverse of differentiation, that is, finding the antiderivative. Besides its apparently greater precision, this definition provided a ready basis for computation of integrals of many functions representable by algebraic expressions.

Nevertheless, it was recognized that an integral could be evaluated, at least approximately, by sums. Indeed, there was a great deal of work on approximating the values of definite integrals—and, analogously, approximating the values of solutions to differential equations—by means of sums, and error bounds were sometimes sought for these approximations. Some of the inequality properties Cauchy needed for the definite integral came from such work.

The study of approximations can lead—and with Cauchy often did lead—to discussions of convergence: approximating the values of infinite series led to treatments of speed of convergence, bounds on error, and ultimately, to the modern theory of convergence; approximating the roots of polynomials led to a proof of existence and a constructive definition of the intermediate value of a continuous function. An analogous development occurred for the integral. Leonhard Euler, Adrien-Marie Legendre, Silvestre-François Lacroix, and Simon-Denis Poisson tried to find bounds on the errors of approximations to the values of definite integrals by means of sums; each tried to show, at least in some special cases, that the error could be made less than any given quantity. Cauchy drew on this work and transformed it into a rigorous theory. The transformation he made was not trivial. Indeed, going from the eighteenth-century treatment of the integral to Cauchy's seems to me to have been the hardest of the steps Cauchy took in rigorizing the calculus.

Why the New Definition?

Why did Cauchy abandon the definition of the definite integral as the antiderivative in favor of its definition as the limit of sums? From a modern viewpoint the answer may seem obvious. Such a definition is absolutely necessary for rigorous calculus. After all, there is no guarantee that every function has an antiderivative. A logically acceptable definition of the integral should support a proof of the existence of the definite integral of a continuous function; defining the integral as a sum does this. But it is not a priori obvious that the limit of the sums approximating an integral always exists, much less that the limit can be proved to exist. What, then, actually led Cauchy to make his definition?

Henri Lebesgue has suggested that pedagogical considerations prompted Cauchy.[4] Lebesgue based this view on the close relationship between Cauchy's definition and the common pedagogical device of approximating a curvilinear area with rectangles. Cauchy himself remarked that "one is naturally led by the theory of quadratures" to consider the integral as a sum.[5] But Euler and Lagrange had found the treatment of the integral as antiderivative sufficiently appealing for textbooks and lectures—presumably because it is easier to explain. And as I shall argue, in fact Cauchy did not derive his sum definition by considering rectilinear approximations to curvilinear areas. Nevertheless, Lebesgue is right in one sense. Cauchy's work on foundations in general, and the theory of the definite integral in particular, was presented in a course of lectures, and lecturing has forced many mathematicians to examine foundations more carefully.

The courses Cauchy gave at the Ecole Polytechnique certainly provided the occasion for his theory of the integral. But the definition itself, as A. P. Iushkevich has said, was chosen by Cauchy to meet "the needs of research."[6] Mathematicians knew many cases in which the definite integral as the area under a curve makes sense, even though the area in question—and therefore the value of the integral—is not simply the antiderivative evaluated at the end points of the interval of integration. Joseph Fourier in particular had exhibited a number of piecewise-continuous functions whose graphs unquestionably enclose areas.[7] His work made clear that the definite integral of a function representable by trigonometric series can exist, even though the function representing the integral is not everywhere differentiable. Following standard practice, Fourier *evaluated*—though he did not *define*—the areas for hard-to-compute integrals as sums.[8] In fact, much simpler counterexamples to the statement "$\int_a^b g'(x)\,dx$ always equals $g(b) - g(a)$" were known. Legendre had treated examples in which definite integrals are taken over intervals containing points at which the functions are discontinuous or become infinite.[9] Cauchy himself had been concerned with this problem in his 1814 memoir on integration.[10] Clearly, some modification of the accepted definition of the definite integral was called for. Cauchy's definition as a sum can be applied to such cases easily, whereas the definition as antiderivative cannot.

Still, this circumstance in itself does not explain why Cauchy sought a new definition. After all, the integral of a piecewise continuous function can be defined simply as the sum of the integrals on the separate pieces; then the integral on each continuous piece can be defined, as before, as the difference between the antiderivatives evaluated at the end points of the intervals of continuity. Points of discontinuity can be handled—and Cauchy had handled them this way in his 1814 memoir on integration—by means of a special formula, the *Cauchy principal value*. In fact, even Cauchy's rigorous definition of 1823 was restricted to continuous functions; again, singular points were handled by the Cauchy principal value.[11]

Difficulties with the integral as antiderivative were not confined to singular points. Even more problems were found in complex integration.[12] Cauchy was quite interested in the topic and had made important contributions to it; as early as the memoir of 1814, he had given a version of the Cauchy–Riemann equations.[13] Indeed, his interest in this subject was a major impetus to his often expressed dissatisfaction with purely formal analogies in mathematical reasoning. He pointed out in the 1814 memoir that one cannot establish theorems in complex analysis simply by carrying over results known for the reals and called instead for "a direct and rigorous analysis."[14] He called reasoning from the real to the complex merely "a kind of induction," and stated explicitly (for the first time) that the integral of a function between two points is the difference between the values of the antiderivatives at those points "only ... in the case of a function which increases or decreases continuously between the limits in question."[15]

Since Cauchy evaluated complex integrals in terms of real integrals, both in his 1814 paper and later, logically he needed a satisfactory theory of the latter to treat rigorously the former. But he did not give it in 1814. Cauchy's 1814 memoir described the integral about a rectangle in terms of the values of the antiderivatives at the end points of straight-line paths;[16] and he wrote this memoir long before he gave his definition of the integral as the limit of sums.

Still, Cauchy's interest in complex integration would have made him dissatisfied with the old definitions of integrals as antiderivatives. Gauss had remarked that if the path of integration between two points includes complex numbers, the value of the integral may depend on the path.[17] Cauchy may not have been acquainted with Gauss's remark; he did know, however, a paper of 1820 by Poisson[18] in which it was observed that the value of an integral like $\int_{-1}^{1} dx/x$ can be different for different paths of integration if one path includes infinity as a value of the function.[19] Thus it no longer could be held that the integral over an interval is simply the difference of the antiderivatives at the end points. Poisson then suggested evaluating such integrals as sums.

Poisson's juxtaposition of a treatment of the integral as a sum with a discussion of the relation between the values of complex integrals and their paths of integration might well have sufficed to influence Cauchy's thinking. The general concept of a line integral obviously clarifies the relation between the value of a complex integral and the path of integration. Only two years after giving his definition of the definite integral as a sum, Cauchy applied the idea to complex integration in a paper of 1825.[20]

In addition Poisson's paper gave a proof that even though the definite integral is not always the difference of the antiderivatives, the integral as sum is still the difference of the antiderivatives at the end points if the function to be integrated is finite throughout the interval. Poisson's proof of this result, which he called "the fundamental proposition of the theory of definite integrals," must have suggested to Cauchy the enormous mathematical fruitfulness of the conception of the integral as the limit of sums—at the same time that Poisson's discussion of complex integrals reemphasized the inadequacy of the old definition of the integral.

I believe that mathematical fruitfulness was the decisive factor in Cauchy's desire for a new definition as well as in his particular choice. I think the main reason he chose to define the integral as the limit of sums was his need to make sure that the object he was defining existed. As Cauchy himself explained,

> In the memoir which has just been read, we consider each definite integral, taken between two given limits, as being nothing else than the sum of the infinitely small values of the differential expression placed under the integral sign, which corresponds to the various values of the variable included between the limits in question. *When this way of viewing definite integrals is adopted, it is easily proved that such an integral has a unique and finite value*, whenever, if the two limits of the variable are finite, the function under the integral sign itself remains finite and continuous throughout the interval included between these limits.[21]

Furthermore, said Cauchy, this definition of the definite integral is "equally suitable to all cases," even to those in which we cannot pass generally from the function under the sign $\int$ to the primitive function.[22] By treating the definite integral as the limit of sums, Cauchy was able to prove the existence of the integral of a continuous function, consider integrals of functions that were not derivatives of known functions, and explain the behavior of integrals along a path. Thus the integral as sum was the answer to many of the perplexing questions raised by the work of Fourier, Gauss, Legendre, Poisson, and—in 1814—Cauchy.

Cauchy's Definite Integral: Definition and Existence

The influence on Cauchy of Euler, Lacroix, and Poisson cannot be denied.[23] Nevertheless, by far the hardest tasks, both technical and conceptual, were accomplished by Cauchy himself. Having returned to the Leibnizian view of the integral as a sum, Cauchy needed to make it more precise. It was not enough to say that the definite integral is the limit of sums. He first had to specify the sums precisely and then had to prove that the limit existed.

He took $f(x)$ to be continuous on a given interval with end points x_0, X and set out to define the definite integral $\int_{x_0}^{X} f(x)\,dx$. He began by dividing the interval into n *not* necessarily equal parts $x_1 - x_0, x_2 - x_1, \ldots, X - x_{n-1}$. He multiplied each of these "elements" by the value of f at its left-hand end point, forming the sum

$$S = (x_1 - x_0)f(x_0) + (x_2 - x_1)f(x_1) + \cdots + (X - x_{n-1})f(x_{n-1}). \tag{6.1}$$

Cauchy noted that the value of S obviously depends on both n and the mode of division of the interval. The crucial question is whether the particular mode of division no longer matters if the size of the subintervals becomes very small and n very large. Using the continuity of $f(x)$ (in his usual stronger version, which we now call uniform continuity), Cauchy was able to prove that the mode of division does not matter, so that S has a unique limit, which he then defined to be the definite integral.[24]

In order to prove that the value of the integral is independent of the mode of dividing up the interval $[x_0, X]$, Cauchy began by choosing the simplest case: there is only one subinterval, $[x_0, X]$ itself. (My translation of Cauchy's definition of the definite integral and the proof of its existence is given in the appendix. Essentially the same proof was given by Moigno in his Cauchy-based text of 1840–1844, *Leçons de calcul différential et de calcul intégral*, vol. 2, pp. 3–6; see G. Birkhoff, *A Source Book in Classical Analysis*, pp. 8–10.) By constructing a partition of this given interval, Cauchy showed that for some constant θ between 0 and 1

$$S = (X - x_0)f[x_0 + \theta(X - x_0)]. \tag{6.2}$$[25]

Returning now to (6.1), Cauchy applied the same technique that produced (6.2) for the one interval to each of the n subintervals $x_1 - x_0$, $x_2 - x_1, \ldots, X - x_{n-1}$ produced by subdividing the original interval. He thus obtained by the same method that for the new subdivision

$$\begin{aligned} S = {} & (x_1 - x_0)f[x_0 + \theta_0(x_1 - x_0)] \\ & + (x_2 - x_1)f[x_1 + \theta_1(x_2 - x_1)] + \cdots \\ & + (X - x_{n-1})f[x_{n-1} + \theta_{n-1}(X - x_{n-1})]. \end{aligned} \tag{6.3}$$

He now defined a set of ε_k values for $k = 0, 1, \ldots, n-1$ by $f[x_k + \theta_k(x_{k+1} - x_k)] = f(x_k) \pm \varepsilon_k$,

so that

$$S = (x_1 - x_0)[f(x_0) \pm \varepsilon_0] + (x_2 - x_1)[f(x_1) \pm \varepsilon_1] + \cdots + (X - x_{n-1})[f(x_{n-1}) \pm \varepsilon_{n-1}]. \tag{6.4}$$

Cauchy then argued that if the subintervals of length $x_k - x_{k-1}$ are taken sufficiently small, then the ε_k will become "very close to zero," so that taking a subpartition of the original partition will not change appreciably the value of S given by (6.1). (This argument actually requires that f be uniformly continuous.)

Cauchy then observed that given any two modes of division whose parts are very small, a third mode of division can always be constructed that subdivides each of the two given ones. The value of S for this new subdivision is arbitrarily close to the value of S for *either* earlier division; and so, Cauchy concluded, as the numerical values of the elements $x_k - x_{k-1}$ become small and n gets very large the different values of S for the two given modes of division differ only imperceptibly from each other. Thus, "If we let the numerical values of these elements decrease while their number increases, the value of S ultimately becomes, for all practical purposes [*sensiblement*] constant. Or, in other words, it ultimately reaches a certain limit. . . ."[26]

Here Cauchy implicitly assumed the same property of the real numbers that he assumed in stating the Cauchy criterion for series:[27] If the various values of some expression—like S in this example, or the nth partial sum of a series—become closer and closer *to each other*, then the expression has a certain limit. In this case, Cauchy remarked, the limit of the value of S depends only on $f(x)$ and the end points x_0 and X of the interval, and "this limit is what is called a definite integral."[28]

Notice again Cauchy's use of an eighteenth-century term—definite integral—in a new and precise way. First he proved that the limit of the expressions given by S existed; only then did he define the integral as equal to that limit. He reintroduced the conventional notation, which he credited to Fourier, $\int_{x_0}^{X} f(x)\,dx$ for the definite integral, but reminded his readers that the symbol $\int$ (a stylized S for *sum*), and the notation $\int f(x)\,dx$, "indicates not a sum of products ... but the limit of a sum of that sort."[29]

Sources for Cauchy's ideas shall include (1) approximations of an integral by sums; (2) detailed discussions of the difference between the approximating sums and the value of the integral; (3) statements that the approximating sums get arbitrarily close to the integral for sufficiently small subdivisions, regardless of the mode of partition; or (4) arguments that a common limit exists for the successive approximating sums (for sufficiently fine common subpartitions). The works of Euler, Lacroix, and Poisson include the first three features; Cauchy's own work (the Cauchy criterion) includes the fourth, so that here Cauchy was his own predecessor.

Euler's Approximation of the Integral by Sums

The first systematic eighteenth-century discussion of approximating the integral as the limit of a sum is found in the *Institutiones calculi integralis* (1768–1770) of Leonhard Euler. Although Euler had defined the integral as the inverse of the differential quotient, he realized that sometimes it was necessary to use other properties of the integral to evaluate it, even approximately. Euler considered X, a function of x whose integral $\int X\,dx$ is to be evaluated between the limits a and x. He divided the interval $[a, x]$ into subintervals of length $a' - a, a'' - a', \ldots, x - {}'x$, where $a < a' < a'' < a''' < \cdots < {}''x < {}'x < x$. To represent the values of the function X at the points $a, a', a'', \ldots, {}'x, x$, he wrote $A, A', A'', \ldots, {}'X$, and X. Finally, he let b be the constant of integration associated with the integral $\int X\,dx$. Euler said that between two values of x differing by a small amount, the function can be treated as a constant, so that the integral $\int X\,dx$ over this small interval is the value of the function there multiplied by the length of the small interval. Thus the value of the integral $\int X\,dx$ between the limits a and x can be approximated by the sum

$$y = b + A(a' - a) + A'(a'' - a') + \cdots + {}'X(x - {}'x).\qquad (6.5)$$[30]

Here, then, is the sum that approximates the value of the definite integral, written (as Cauchy later wrote it) with the function evaluated at the left-hand end points of the subintervals.

Euler did not compute a general error estimate for the

approximation (6.5). He did discuss, however, different ways in which the approximation's closeness could be ensured. First, he said, assume that the function is either always increasing or always decreasing on the given interval. Then the integral $y = \int X\,dx$ between the given limits is always included between the sum of the lengths of the subintervals times the values of X taken at the left-hand end points of the subintervals, and the corresponding sum for the values of X taken at the right-hand end points. This gives an explicit error bound for the integral of a monotonic function.[31]

If the function is not always increasing or decreasing, this error bound will not work. Nevertheless, Euler still argued that the approximation could be made closer and closer to the true value by carefully choosing the subintervals; if the value of the function between two given values of x changes violently, very small intervals must be taken; if the value of the function between two given values of x does not change violently, large intervals may be taken. Thus for Euler the variation of the function determines whether the given interval should be divided into equal or unequal, large or small subintervals.[32]

Euler and Cauchy used the same kinds of approximating sums and realized that finer subdivisions produce better approximations, but their views of the integral differed greatly. Euler believed that the integral had an existence independent of the approximation procedure. Further, he could not prove the accuracy of the approximation for a function that is not piecewise monotonic. In this is seen the characteristic attitude of the eighteenth-century approximator: The thing exists, and our job is to approximate it in the most expeditious way. Not until the work of Cauchy was the general question of the existence of the integral even raised, much less rigorously answered.

Meanwhile, Euler's approximations proved fruitful, in both his hands and the hands of others. For example, Adrien-Marie Legendre (1752–1833) used the sum techniques to approximate the values of integrals over intervals including points at which the function is infinite. He also gave an estimate of the "error" in the usual integral formula introduced by this singular point.[33] Several times in his *Analytical Theory of Heat*, Fourier used the principle that an integral could be treated as a sum.[34] Fourier's work, like Legendre's and Euler's, showed the usefulness of considering the integral as a sum when computation of the antiderivative failed to give the integral's exact value. Interesting though the works of Legendre and Fourier may be, however, they do not seem to have been decisive in determining Cauchy's approach to these problems.

Just as Lacroix's *Traité du calcul* had transmitted to Cauchy eighteenth-century work on limits and convergence, most probably it played the major role in transmitting to him also Euler's approximations to the integral. As usual, Lacroix had not intended to do anything new; in elaborating Euler's work, his goal was to present, explain, and clarify. But whatever the modesty of Lacroix's intentions, his description of the Euler integral made the approximating formulas for the integral as sum readily available to Cauchy.

Lacroix followed Euler closely in stating that the definite integral of the function X on an interval $[a, a_n]$ is approximated by the sum of the values of X evaluated at the left-hand end points of the subintervals into which the given interval is divided. That is, if the given interval is $[a, a_n]$, $a < a_1 < a_2 < \cdots < a_n$, Y is a constant of integration, and $Y' = X(a)$, $Y_1' = X(a_1), \ldots, Y_{n-1}' = X(a_{n-1})$, the integral may be approximated by the sum

$$Y + Y'(a_1 - a) + Y_1'(a_2 - a_1) + \cdots + Y'_{n-1}(a_n - a_{n-1}).\text{[35]} \tag{6.6}$$

Lacroix went beyond Euler in the discussion of the closeness of the approximation to the true value, perhaps because he was able to draw on more such work in 1797 than Euler in 1768, or perhaps because he had the pedagogically valuable desire to explain things at length. Like Cauchy, Lacroix began with just one interval. He took the interval $[a, a_1]$ and let the function X be always increasing or always decreasing on that interval.[36] Then, as Euler had shown, Lacroix said that the integral is bounded by the approximations $Y + Y'(a_1 - a)$ and $Y + Y_1'(a_1 - a)$.[37] Lacroix then further subdivided that same interval from a to a_1 by means of the intermediate points $\alpha_1, \alpha_2, \alpha_3, \ldots, \alpha_m$, so that $a < \alpha_1 < \alpha_2 < \cdots < \alpha_m < a_1$. Thus the integral lies between

(6.7a) $$Y + Y'(\alpha_1 - a) + Y'(\alpha_2 - \alpha_1) + \cdots + Y'(a_1 - \alpha_m)$$

and

(6.7b) $$Y + Y_1'(\alpha_1 - a) + Y_1'(\alpha_2 - \alpha_1) + \cdots + Y_1'(a_1 - \alpha_m).$$

But, because the function X was assumed always increasing or decreasing, $X(\alpha_1), X(\alpha_2), \ldots, X(\alpha_m)$ all lie between $X(a)$ and $X(a_1)$. Let $X(\alpha_1) = y_1'$, $X(\alpha_2) = y_2', \ldots, X(\alpha_m) = y_m'$. Then the expression

(6.8) $$Y + Y'(\alpha_1 - a) + y_1'(\alpha_2 - \alpha_1) + y_2'(\alpha_3 - \alpha_2) + \cdots + y_m'(a_1 - \alpha_m)$$

must lie between (6.7a) and (6.7b).[38] The partition by the α_k in (6.8) is a subpartition of the original partition; and since (6.8) lies between expressions (6.7a) and (6.7b) bounding the integral, (6.8) itself must be close to the integral. In fact, said Lacroix, (6.8) can be made as close to the integral as desired if the points $\alpha_1, \alpha_2, \alpha_3, \ldots, \alpha_m$ are taken sufficiently close together; in his words, "by imagining a sufficient number of terms" (6.8) can become as close as desired to the "true value."[39] (Lacroix seems to have assumed that simply increasing the number of terms α_k between a and a_1 suffices to make (6.8) approach the integral as closely as possible. But, since the α_k are not equally spaced, he should have required also that *each* $\alpha_k - \alpha_{k-1}$ become arbitrarily small.)

Of course, the result would not have surprised Euler. Indeed, except for the analytic notation the result would not have surprised Archimedes. But it is the specific *form* of Lacroix's result that is of interest, in particular as Cauchy's most likely immediate source. Cauchy had read the *Traité*, including the section on integration, and consistently used Lacroix's terminology—*element* for subinterval, *arbitrary constant*, and above all *definite integral* and *indefinite integral*. Cauchy himself acknowledged the kinship between his sums and earlier approximations by pointing out that his formula (6.1) and the corresponding formula evaluating the function at the right-hand end points, "are frequently employed in finding approximate values of definite integrals."[40] And Cauchy's *Calcul infinitésimal* includes a number of approximations just like those given by Lacroix.[41]

What might Cauchy have owed to Lacroix's *Traité du calcul*? Lacroix had picked out the key property of the definite integral—the integral is the limit of sums—and used it in a proof. Moreover, not only had Lacroix shown that the integral of a monotonic function is included between two computable finite bounds, but he had spelled out in an algebraic argument that finer subdivisions can be shown to produce closer approximations. He also had stated that the interval over which a function is evaluated may be broken up into parts in such a way that in each part the function increases or decreases;[42] thus the integral of the function can be divided into a sum of integrals in such a way that each can be given bounds. Lacroix's work thus implies, though not saying explicitly, that for any piecewise monotonic function approximating sums can be found that are arbitrarily close to the function's integral. These insights, from whatever source Cauchy may have first gleaned them, were necessary to his theory of the integral.

The Existence of the Integral and Its Independence from the Mode of Division: Poisson and Cauchy

Still, Lacroix cannot be viewed as the predecessor for Cauchy's proof of the existence of the definite integral. Perhaps the approximation $\sum_{k=0}^{n-1} f(x_k)(x_{k+1} - x_k)$ or simply the corresponding geometric picture might have suggested that the closeness of the approximation can be shown, at least in some cases, to increase with the fineness of the partition. And certainly the possibility of changing once again an approximation process into an existence proof might have occurred to Cauchy upon reading either Euler's or Lacroix's work. Nevertheless, Euler and Lacroix did not try to prove that the true value of the integral of an arbitrary function differs from the approximating sums by less than any given quantity for sufficiently small subintervals. Nor did Lacroix explicitly state, let alone prove, that expressions like his sum (6.8) have as their limits values independent of the particular choice of the subdividing points $\alpha_1, \alpha_2, \ldots, \alpha_m$ provided the α_k are sufficiently close. Simon-Denis Poisson, however, did address himself to these problems.

Poisson's work appeared in the 1820 paper dealing with aspects of real and complex integration.[43] Poisson had tried to prove that if the function to be integrated on an interval is finite there, then the integral evaluated as a sum is equal to the difference of the antiderivatives.[44] Appropriately, he called this result "the fundamental proposition of the theory of definite integrals." His result is not quite our fundamental theorem of calculus, since it begs the question of the existence of both the definite and indefinite integral, and only works for functions that are antiderivatives. Also, his proof is limited to the case in which the subintervals are equal. Still, the importance of his result is obvious.

Poisson took the function $f(x)$, finite from $x=a$ to $x=b$, and designated its antiderivative by $F(x)$. He reminded the reader that $F(b) - F(a)$ "is what is called the definite integral" of $f(x)$ from $x = a$ to $x = b$.[45] After dividing the given interval $[a,b]$ into n parts of length α, Poisson undertook to prove that if n becomes large and the sum S is defined by

$$S = \alpha[f(a) + f(a + \alpha) + \cdots + f\{a + (n-1)\alpha\}], \tag{6.9}$$

then, "$F(b) - F(a)$ is exactly [*rigorousement*] the limit of the sum represented by S."[46]

To prove that, as n increases, the limit of the sum S is equal to the integral, Poisson made use of an approximation to the value of an integral not yet discussed. This approximation, like so many others, was described in detail first by Euler. Euler had recognized that the contribution to the integral of each term in a sum like (6.9)—for instance, the term $\alpha f(a)$—is only approximate, since it treats the function f in the subinterval $[a, a + \alpha]$ as though it is the constant $f(a)$. A closer value for the contribution to the integral of the first subinterval can be given according to Euler by the Taylor series for $F(a + \alpha) - F(a)$, where

$$F(a + \alpha) = F(a) + \alpha F'(a) + \alpha^2/2!\, F''(a) + \cdots.$$[47]

Analogous expressions can be written instead of $f(a + \alpha)$, $f(a + 2\alpha), \ldots.$

Poisson took these approximations and supplied the Taylor series with their remainders. Taylor's theorem gave Poisson for the function F

$$F(y + z) = F(y) + zf(y) + Rz^{1+k},* \tag{6.10}$$

where y is the independent variable and z its increment, $f(x) = F'(x)$, R is the remainder (assumed finite), and k is positive.[48]
Thus, he obtained

$$F(a + \alpha) = F(a) + \alpha f(a) + R_0\alpha^{1+k},$$
$$F(a + 2\alpha) = F(a + \alpha) + \alpha f(a + \alpha) + R_1\alpha^{1+k},$$
$$\vdots$$
$$F(a + n\alpha) = F[a + (n - 1)\alpha] + \alpha f[a + (n - 1)\alpha] + R_{n-1}\alpha^{1+k}.$$

Combining these equations and noting that $a + n\alpha = b$ by definition of α,

$$F(b) = F(a) + \alpha f(a) + \alpha(a + \alpha) + \cdots + \alpha f[a + (n - 1)\alpha] + (R_0 + R_1 + \cdots + R_{n-1})\alpha^{1+k}.$$

Poisson now set M to be the maximum value over the interval $[a, b]$ of the remainder function, "abstraction made of the sign." Then $(R_0 + R_1 + \cdots + R_{n-1}) \leqslant nM$. Recalling the definition of S in (6.9) and the fact that $n\alpha = b - a$, he then obtained $|F(b) - F(a) - S| \leqslant |nM(\alpha^{1+k})|$, and thus

* The Lagrange form is not used for R, nor is the exact form of R important as long as it is finite. Poisson's knowledge of Lagrange is attested by a reference in "Suite du mémoire" (p. 320) to Lagrange's *Calcul des fonctions*. Lacroix had not supplied a remainder in this approximation, but instead had used "Euler's criterion" to argue that for sufficiently small values of the increment (for which Poisson wrote z), $zf(y)$ exceeds the sum of all the rest of the terms (Lacroix, *Traité du calcul*, 1st ed., vol. 2, p. 136). Poisson knew Lacroix's treatises; see, for example, the explicit reference in "Suite du mémoire" (p. 319) to the treatment of approximating definite integrals as sums in section 471 of the second edition of the *Traité du calcul*.

In discussing (6.10), Poisson stated that if the second differential of F is infinite at a few points, the expressions are still valid and the third term of the Taylor series development of $F(y + z)$ at those points "will contain a power of z whose exponent lies between 1 and 2." Thus he wrote Rz^{1+k}, and not Rz^2 ("Suite du mémoire," p. 322).

(6.11) $$|F(b) - F(a) - S| \leqslant |M(b-a)\alpha^k|.$$[49]

As n increases, α decreases; thus since k is positive, the left-hand side of the inequality (6.11) "can become less than any given quantity." Poisson concluded that "$F(b) - F(a)$ is exactly the limit of the sum represented by S."[50] (The form of the bound $M(b-a)\alpha^k$ may have been suggested by either the Lagrange or the Lacroix notation for the mean-value theorem for integrals.[51])

Poisson's proof is not bad. His theorem really means that the difference of the antiderivatives of f, namely, $F(b) - F(a)$, can be approximated by the sums (6.9) to within any given quantity. Poisson's proof easily can be made acceptable by modern standards if F is assumed to have bounded second derivatives, that is, if f is assumed to have antiderivatives and bounded first derivatives, on the given interval. Poisson's proof well may have suggested to Cauchy that the equivalence of the conceptions of the integral as antiderivative and as sum could be proved—at the same time that Poisson's remarks about complex integrals made clear the general inadequacy of the antiderivative as a definition of the integral.

One major weakness of Poisson's proof is that it is restricted to the case of equal divisions of the interval. Poisson recognized that he had to justify the conclusion of his proof for the case of unequal divisions as well. But he could not demonstrate with comparable rigor that the value of the integral did not depend on the mode of division for this case. He simply made an appeal to geometry of the kind that Lagrange and Bolzano were always warning against: "If the integral is represented by the area of a curve, *this area will be the same*, if we divide the difference of the extreme abscissas into an infinite number of equal parts, or an infinite number of unequal parts following any law."[52] Poisson's statement must have cried out to Cauchy for proof. And Cauchy proved it.

The Fundamental Theorem of Calculus

Once Cauchy had defined the integral and established its existence, he was able to prove theorems about it.. In particular, he was able to obtain what is now called the fundamental theorem of calculus: If $f(x)$ is finite and continuous throughout the interval $[x_0, X]$ and

$$\mathscr{F}(x) = \int_{x_0}^{X} f(x)\,dx,$$

then

$$\mathscr{F}'(x) = f(x).$$[53]

Ever since Newton and Leibniz, the fundamental theorem of calculus has been the key result of the calculus. For Cauchy and his successors, it relates his new rigorous definitions of derivative and integral by means of one theorem linking the differential calculus with the integral calculus. Though Cauchy himself did not call the theorem fundamental, the place it occupies in the calculus amply justifies this designation.

Cauchy proved the fundamental theorem by combining two things: the mean-value theorem for integrals,[54] already known to Lagrange;[55] and the additivity of the definite integral over intervals, a fact long known and, for Cauchy, an easy consequence of the sum definition of the integral. (The proof—which is closely akin to that used today—is given in translation in the appendix.) The key step is

$$\mathscr{F}(x+\alpha) - \mathscr{F}(x) = \int_{x_0}^{x+\alpha} f(x)\,dx - \int_{x_0}^{x} f(x)\,dx$$

$$= \int_{x}^{x+\alpha} f(x)\,dx = \alpha f(x+\theta\alpha)$$

where $0 \leqslant \theta \leqslant 1$. The theorem now follows from Cauchy's definition of the derivative.

The sources of Cauchy's proof are twofold: Lagrange's proof of what amounts to the fundamental theorem, in which there is a careful consideration of expressions like $\mathscr{F}(x+\alpha) - \mathscr{F}(x)$; and Lagrange and Lacroix's statements of the mean-value theorem for integrals (though neither fully appreciated the importance of the result).

Lagrange's Proof of the Fundamental Theorem

How, one might ask, could Lagrange prove the fundamental theorem of calculus? He had no definition of the definite integral; to him, the indefinite integral was just the antiderivative. Yet he did prove a version of the theorem: Lagrange proved that $f(x) = F'(x)$ for a function $F(x)$ defined as the area under the curve $y = f(x)$ up to some x. In fact, Lagrange's proof of the theorem is technically very much like Cauchy's, even though the logical settings of the two proofs are entirely different.

The area under the curve $y = f(x)$ from x to $x + i$ is given, said Lagrange, by $F(x + i) - F(x)$, where F represents the area function. Let the function f be increasing on the interval $[x, x + i]$. In this case, the geometry of the situation makes it clear that the area under the portion of the curve between x and $x + i$ lies between $if(x)$ and $if(x + i)$. That is,

$$if(x) \leqslant F(x+i) - F(x) \leqslant if(x+i).\text{[56]} \tag{6.12}$$

Of course the result is obvious. But, Lagrange continued, the Taylor series with remainder implies that

$$f(x+i) = f(x) + if'(x + j_f),$$
$$F(x+i) = F(x) + iF'(x) + (i^2/2)F''(x + j_F).$$

(Lagrange used only j for both the quantities I have called j_f and j_F, but recognized *explicitly* that they are distinct.) Thus, (6.12) now becomes

$$if(x) \leqslant iF'(x) + (i^2/2)F''(x + j_F) \leqslant if(x) + i^2f'(x + j_f).$$

Therefore

$$0 \leqslant i[F'(x) - f(x)] + (i^2/2)[F''(x + j_F)] \leqslant i^2f'(x + j_f). \tag{6.13}$$

(6.13) holds for all i, no matter how small. Therefore, said Lagrange, $F'(x) - f(x)$ must be zero. If not, he pointed out, $|i| < |[F'(x) - f(x)]/[f'(x + j_f) - (1/2)\ F''(x + j_F)]|$ would make (6.13) false.[57] Therefore, it must follow that $F'(x) = f(x)$.

Lagrange's Taylor-series expressions play the same logical role in his proof that the mean-value theorem for derivatives plays in Cauchy's. Furthermore, Lagrange carried out his proof of the theorem with a sophisticated treatment of the relevant inequalities. These resemblances are sufficiently close to suggest influence—possibly direct, possibly through the work of Poisson (which obviously is based on Lagrange's work)—on Cauchy.

But Lagrange's proof has limitations deriving from its context, if not its internal logic. Lagrange assumed—as Poisson, following in his footsteps, did in his 1820 paper—that the function F had two derivatives: that is, he explicitly used the assumption that f had both derivatives and antiderivatives. Moreover, he had defined neither area nor integral, and his proof requires that the function be monotonic. Thus he did not have the essence of Cauchy's proof of the fundamental theorem.

To adapt Lagrange's proof for his own purposes, Cauchy had to define and to prove the existence of F. He did so. In addition he had to find and prove a result equivalent to (6.12), but not limited to monotonic functions. This he did by stating the mean-value theorem for definite integrals. The mean-value theorem for integrals had been derived by both Lagrange and Lacroix.[58] Each had derived it from Lagrange's lemma that a function with positive derivative on an interval is increasing there. But each had defined the integral as the inverse of the derivative and thus considered this theorem as a variant of the mean-value theorem for derivatives. Because Cauchy saw the integral as a sum, not an antiderivative, he needed a new proof of the mean-value theorem for integrals. The mean-value theorem for integrals was stated by Cauchy thus:

$$\int_{x_0}^{X} f(x)\,dx = (X - x_0) f[x_0 + \theta(X - x_0)],\ 0 \leqslant \theta \leqslant 1. \tag{6.14}$$

He derived this theorem from one of the steps in his proof that the mode of division does not affect the value of the definite integral. Once again Cauchy had taken an earlier result, given it a different logical basis, and used it for an entirely different purpose. Based on an acceptable proof of the mean-value theorem (Cauchy's does not meet modern standards without modification), Cauchy's proof of the fundamental theorem can be used today.

Cauchy's Theory of the Integral and His Predecessors

The works of Euler, Lagrange, Lacroix, and Poisson contain useful techniques, fundamental questions, and a wealth of examples whose direct bearing on Cauchy's formulation of the essential properties of the definite integral can be documented.

Cauchy acknowledged that he had read the textbooks of Euler.[59] Also, Cauchy's proof of the existence of the solution to a differential equation by what is now called the Cauchy–Lipschitz method is based on a method of constructing the approximate solution first explained in Euler's *Institutiones calculi integralis*.[60] Indeed, the relationship between Cauchy's existence proof and Euler's approximation is exactly analogous to the case of the definite integral.

I have already pointed out that Cauchy acknowledged Poisson's 1820 paper several times. Moreover, Poisson and Cauchy were colleagues at the Ecole Polytechnique and on occasion must have discussed areas of common concern, among which surely would have been integrals. Numerous resemblances between their works show that Cauchy knew and used Lacroix's books. Finally, I have given evidence several times for Cauchy's acquaintance with Lagrange's *Fonctions analytiques*.

But the technical similarities in their treatments of the definite integral cannot dispel the differences in points of view between Cauchy and his predecessors. For Euler and Lacroix, approximation by sums is just one property of the integral, related to little else in the theory of the integral calculus. For Cauchy, it became the fundamental and defining property. For Euler and Lacroix, the integral is the antiderivative, whose value can be approximated by sums. For Cauchy, the integral is the limit of sums of a certain type—a limit whose existence has to be shown: If the integral $\int_a^b f(x)\,dx$ exists and equals $F(b) - F(a)$, where $F'(x) = f(x)$, then these facts have to be proved, not merely stated.

Similar comments apply to the work of Poisson. In fact, his work serves as a reminder that merely describing the work before Cauchy cannot explain Cauchy's accomplishment. Poisson had the same materials at hand in 1820 as Cauchy in 1823: the work of Lacroix and Euler on approximating integrals as sums; the Taylor-series approximations to the integral; Lagrange's use of Taylor series with remainder; the mean-value theorem for integrals; and of course the relationship between the integral and the antiderivative. But this common historical background was not enough for Poisson. Nor does it help much to point out that Cauchy was required to teach courses at the Ecole Polytechnique; Poisson was too.

In the case of integrals even more than in the cases of limits, continuity, convergence, and derivatives, the achievements of Cauchy's predecessors, though necessary, were far from sufficient. None of his predecessors even appreciated the need for a rigorous theory of the integral. Only one man had the genius to choose—and to choose almost unerringly—the techniques that would be fruitful, put those techniques together into proofs, and state and prove the additional results required.

Applications and Influence of Cauchy's Theory of the Integral

The second part of Cauchy's *Leçons sur les applications du calcul infinitésimal à la géometrie* includes applications of his theory of the integral to geometry. Some of the applications—for example, to arc lengths or the calculation of specific areas or volumes—are done much as they had been done in the eighteenth century. But to show that the definite integral represents the area under a curve, Cauchy used the mean-value property of areas and his mean-value property of derivative and integral—an approach vaguely akin to Lagrange's derivation of his version of the fundamental theorem of calculus.[61] Cauchy used analogous considerations to derive the integrals for the areas between two plane curves[62] and to find the area of a curved surface[63] and the volumes of solids.[64]

One motivation for Cauchy's theory of the integral was his perception that such a theory would be needed to study complex integrals. And in fact Cauchy did apply his definition of the integral as the limit of sums from the 1823 *Calcul infinitésimal* to the 1825 paper "Mémoire sur les intégrales définies prises entre des limites imaginares" (first published in 1874–1875), in which appear the Cauchy integral theorem and the calculus of residues. The introduction to this 1825 memoir emphasized the principal value of improper integrals rather than the integral as the limit of sums in the continuous case. Nevertheless, Cauchy referred there to recent work of his showing how to "fix in every possible case" the sense of the notation $\int_{x_0}^{X} f(x)\, dx$. He then said, "I propose today to apply the principles which have guided me in those researches to integrals taken between imaginary [that is, complex] limits."[65]

Cauchy began by recalling the definition of the real definite integral as the limit of sums. Then, "to include integrals taken between real limits and imaginary limits in the same definition," he defined the complex integral

$$\int_{x_0 + y_0\sqrt{-1}}^{X + Y\sqrt{-1}} f(z)\, dz \text{ (where } z = x + y\sqrt{-1})$$

as "the limit, or one of the limits" toward which the sums (in modern notation)

$$\sum_{j=1}^{n} f(x_j + iy_j)[(x_{j+1} - x_j) + i(y_{j+1} - y_j)]$$

converge when the terms of the monotonic sequences $x_0, x_1, \ldots, x_{n-1}, X$ and $y_0, y_1, \ldots, y_{n-1}, Y$ "approach each other indefinitely as their number increases."[66] Cauchy used his definition to show that when the sequences are chosen from the curve $x = \phi(t)$ and $y = \chi(t)$, the complex integral is given by

$$(6.15)\quad \int_{x_0 + y_0\sqrt{-1}}^{X + Y\sqrt{-1}} f(z)\, dz = A + B\sqrt{-1}$$

$$= \int_{t_0}^{T} (\phi'(t) + \chi'(t)\sqrt{-1}) f(\phi(t) + \chi(t)\sqrt{-1})\, dt.\text{[67]}$$

Cauchy then showed by using ideas from the calculus of variations that if "the function $f(x + y\sqrt{-1})$ remains finite and continuous," then the "expression $A + B\sqrt{-1}$ is independent of the nature of the functions $x = \phi(t)$ and $y = \chi(t)$."[68] This result is often called the Cauchy integral theorem. (Cauchy's proof of the integral theorem and other results in this memoir needs more than the continuity of f. In fact his proofs always assume that f has as many complex derivatives as needed.) Then he used the integral theorem and his theory of principal values of improper integrals to prove the Cauchy integral formula and develop the calculus of residues, which in turn he applied to the evaluation of improper integrals.

To what extent did Cauchy actually need the 1823 sum definition of the definite integral to do the 1825 work on complex integration? The definition's role seems to have been purely foundational; it does not enter into the proofs of any of the theorems once the parameterized form (6.15) has been obtained. Of course Cauchy could not define the complex integral as an antiderivative; yet it had to be introduced somehow. But had the 1825 paper begun with a definition of the complex integral as satisfying the parameterized equation (6.15), the rest of the paper could have proceeded unchanged, largely on the basis of facts about integrals already accepted. Cauchy, however, wanted a rigorous foundation for his theory of complex integration, and the basis he chose was his definition of real-valued integrals. Indeed, because the parameterized equation followed from the sum definition in the "direct and rigorous" manner he so prized, Cauchy felt free to let it serve—just as the Lagrange property of the derivative had served—as the basis for an entire topic. Thus although the sum definition was not built into his proofs about complex integrals as it had been into his proof of the existence of the real definite integral, his use of it shows both that complex integration partly motivated his definition and that it was one area of the definition's application.

Any assessment of Cauchy's definition of the integral must conclude with the observation that Bernhard Riemann drew on it to develop his own theory of the integral. Of course Riemann's theory differs from—and goes beyond—Cauchy's. First, Cauchy took the values of f at the left-hand end points of subintervals; Riemann took any arbitrary point in his subintervals. (But Cauchy's use of the expression $S = \Sigma_k f[x_k + \theta_k(x_k - x_{k-1})]$, $1 \geqslant \theta_k \geqslant 0$, to represent S for another mode of division shows that for continuous functions at least, evaluation at the left-hand end point is not essential.) Second, and more fundamental, Cauchy assumed explicitly that the function whose integral is to be defined is continuous, while implicitly assuming that it is uniformly continuous; Riemann assumed neither and in fact gave an example of an integrable function with infinitely many discontinuities in an arbitrarily small interval. Riemann stated that when the limit of his sum expressions exists, it is the definite integral; when the limit does not exist, the expression $\int_a^b f(x)\,dx$ has no meaning. Thus Riemann extended Cauchy's definition to a wider class of functions. (For continuous functions, Cauchy's and Riemann's integrals have the same value. Note that Cauchy extended his conception of the integral to include functions with finitely many discontinuities, that is, improper integrals, as early as 1814.[69]) But the Riemann integral is a natural development of Cauchy's ideas, in the same way as are late-nineteenth-century theories of continuity, convergence, and derivatives.[70]

Conclusion

What is our estimate of Cauchy's achievement? Cauchy's work established a new way of looking at the concepts of the calculus. As a result, the subject was transformed from a collection of powerful methods and useful results into a mathematical discipline based on clear definitions and rigorous proofs. His views were less intuitive than the old ones, but they provide a new set of interesting questions. His definition of limit and elaboration of the associated method of proof by inequalities are the basis for modern theories of continuity, convergence, derivative, and the integral. And many of the important consequences of these theories—in the study of convergence, existence proofs for the solution of differential equations, and the properties of definite integrals—were pioneered by Cauchy himself.

Moreover, Cauchy's rigorization of the calculus was much more than the sum of its separate parts. It was not merely that Cauchy gave this or that definition, proved particular existence theorems, or even presented the first reasonably acceptable proof of the fundamental theorem of calculus. He brought all these things together into a logically connected system of definitions, theorems, and proofs.

The implications of this achievement go beyond the calculus. In a very important sense, it may be said that Cauchy brought ancient and modern mathematics together. He cast his rigorous calculus in the deductive mold characteristic of ancient geometry. And unlike his predecessors, he did this successfully; that is, he, not only gave his work a Euclidean form but presented definitions that generally are adequate to support the desired results, proofs that basically are valid, and methods that were fruitful sources for later mathematical work. Cauchy, then, brought together three elements: the major *results* of analysis, most of which he could now prove; some fruitful *concepts* and *techniques* from algebra (particularly algebraic approximations) and analysis; and the *rigor* and *proof structure* of

Greek geometry. For a long time Greek geometry had been considered the model for all of mathematics. If the origins of modern mathematics are traced to the Renaissance, then the rigor and structure characteristic of Greek geometry first effectively became part of modern mathematics only with Cauchy's work. Of course the late-nineteenth-century idea that mathematics is the science of abstract logical systems in general is absent from Cauchy's work. But Cauchy's rigorization of the calculus was an indispensable first step in that direction.

Cauchy left some unfinished business, as subsequent history shows. Some gaps in specific proofs had to be filled; some assumptions had to be proved, or at least explicitly stated; some crucial distinctions had yet to be made. But there is little in nineteenth-century analysis that was not marked, directly or indirectly, by his ideas. The basic logical structure Cauchy erected provides the framework in which we still think about rigorous calculus.

Tracing Cauchy's innovations in the foundations of analysis back to their sources yields a pleasing new result for the historian. Historians looking at the eighteenth-century debates over the foundations of the calculus had seen comparative mathematical unknowns—Bishop Berkeley, James Jurin, Benjamin Robins, Simon L'Huilier, and Lazare Carnot—among Cauchy's most important predecessors. To be sure, their philosophical discussions, as well as those of Maclaurin, Euler, D'Alembert, and Lagrange, influenced Cauchy. But Cauchy needed more than philosophical discussions; he needed mathematics. And the mathematics he needed came from the work of the major mathematicians of the eighteenth century. I find this conclusion satisfying because it seems to reflect more convincingly than older views the way in which one expects great mathematicians to relate to their greatest predecessors. The evidence shows that the men to whom Cauchy owed most in working out his foundations were Euler, D'Alembert, Ampère, Poisson, and—above all—Lagrange, and this seems altogether fitting.

Cauchy's originality in the foundations of the calculus lies in part in the use he made of what others had done. Yet, as the work of many historians of scientific ideas reminds us, there can be as much innovation in transforming old methods as in developing new ones. The fact that many of the basic techniques of Cauchy's calculus existed in the eighteenth century should increase, not decrease, our wonder at his achievement. Cauchy was able to see—where nobody else had been able to see—how these ideas could be used to build a new rigorous calculus. We do not insist that an architect make every brick he uses with his own hands; instead, we marvel that the beauty of his creations can come from such commonplace materials. Augustin-Louis Cauchy neither began nor completed the rigorization of analysis. But more than any other mathematician, he was responsible for the first great revolution in mathematical rigor since the time of the ancient Greeks.

Appendix: Translations from Cauchy's Oeuvres

Proof of the Intermediate-Value Theorem for Continuous Functions: *Cours d'analyse*, 1821, Note III (*Oeuvres*, Series 2, Vol. 3, pp. 378–380)

Theorem. Let $f(x)$ be a real function of the variable x, continuous with respect to that variable between $x = x_0$, $x = X$. If the two quantities $f(x_0)$, $f(X)$ have opposite sign, the equation

(1) $f(x) = 0$

can be satisfied by one or more real values of x between x_0 and X.

Proof. Let x_0 be the smaller of the two quantities x_0, X. We will set $X - x_0 = h$, and we designate by m any integer greater than one. Because one of the two quantities $f(x_0)$, $f(X)$ is positive, the other negative, it follows that if we form the sequence

$$f(x_0), f(x_0 + h/m), f(x_0 + 2h/m), \ldots, f(X - h/m), f(X),$$

and compare successively the first term in that sequence with the second, the second with the third, the third with the fourth, etc., we necessarily will finish by finding once—or more than once—two consecutive terms that are of opposite sign.

Let $f(x_1)$, $f(X')$ be two such terms, where x_1 is the smaller of the two corresponding values of x. Clearly we will have

$$x_0 < x_1 < X' < X$$

and

$$X' - x_1 = h/m = 1/m\,(X - x_0).$$

[Cauchy uses the sign $<$ for *less than or equal to*.] Having determined x_1 and X' in the way just described, similarly we can place between these two new values of x two other values x_2, X'', which, when substituted in $f(x)$, give results of opposite sign, and which satisfy the conditions

$$x_1 < x_2 < X'' < X'$$

and

$$X'' - x_2 = 1/m(X' - x_1) = 1/m^2(X - x_0).$$

Continuing thus, we obtain, first, a series of increasing values of x, that is,

(2) $x_0, x_1, x_2, \ldots;$

second, a series of decreasing values

(3) $X, X', X'', \ldots$

whose terms, since they exceed those of the first series by quantities equal, respectively, to the products

$$(1) \cdot (X - x_0),\ (1/m)(X - x_0),\ (1/m^2)(X - x_0), \ldots,$$

ultimately will differ from the values in the first series by as little as desired. We must conclude from this that the general terms of the series (2) and (3) converge toward a common limit. Let a be that limit. Since the function $f(x)$ is continuous between $x = x_0$ and $x = X$, the general terms of the following series,

$$f(x_0), f(x_1), f(x_2), \ldots, f(X), f(X'), f(X''), \ldots$$

both converge toward the common limit $f(a)$; and since they always remain of opposite sign when they approach this limit, it is clear that the quantity $f(a)$, which must be finite, cannot differ from zero. Therefore, the equation

(1) $f(x) = 0$

will be satisfied if we give the variable x the particular value a, which lies between x_0 and X. In other words,

(4) $x = a$ will be a *root* of equation (1).

The First Rigorous Proof about Derivatives: *Calcul infinitésimal*, Leçon 7 (*Oeuvres*, Series 2, Vol. 4, pp. 44–45)

Theorem. If the function $f(x)$ is continuous between the limits [that is, bounds or end points] $x = x_0, x = X$, and if we let A be the smallest, B the largest value of the derivative $f'(x)$ in that interval, the ratio of the finite differences

(4) $[f(X) - f(x_0)]/(X - x_0)$

must be included* between A and B.

* I have translated Cauchy's *comprise* as "included" and his *renfermé* as "lying between." The context of the proof makes clear that c is included between a and b means $a \leqslant c \leqslant b$ and that c lies between a and b means $a < c < b$.

Proof. Let δ, ε be two very small numbers; the first is chosen so that for all numerical [that is, absolute] values of i less than δ, and for any value of x included between the limits x_0, X, the ratio

$$f(x+i) - f(x)/i$$

will always be greater than $f'(x) - \varepsilon$ [Cauchy's *Oeuvres* has the misprint $f(x) - \varepsilon$] and less than $f'(x) + \varepsilon$. If we interpose $n - 1$ new values of the variable x between the limits x_0, X, that is,

$$x_1, x_2, \ldots, x_{n-1},$$

so that the difference $X - x_0$ is divided into elements

$$x_1 - x_0, x_2 - x_1, \ldots, X - x_{n-1}$$

that all have the same sign and numerical values less than δ, then, since of the fractions

$$(5)\quad f(x_1) - f(x_0)/x_1 - x_0, f(x_2) - f(x_1)/x_2 - x_1, \ldots,\\ f(X) - f(x_{n-1})/X - x_{n-1}$$

the first will be included between the limits $f'(x_0) - \varepsilon$, $f'(x_0) + \varepsilon$, the second between the limits $f'(x_1) - \varepsilon$, $f'(x_1) + \varepsilon, \ldots,$ etc., each of the fractions will be greater than $A - \varepsilon$ and less than $B + \varepsilon$. Moreover, since the fractions (5) have denominators of the same sign, if we divide the sum of their numerators by the sum of their denominators, we obtain a *mean* fraction, that is, one included between the smallest and the largest of those under consideration (see *Analyse algébrique*, Note II, Theorem XII).*

Thus the expression (4), with which that mean coin-

* Here Cauchy was referring to this theorem: "If $b, b', b'', \ldots$ are n quantities with the same sign, and if $a, a', a'', \ldots$ are any n quantities, we have $\dfrac{a + a' + a'' + \cdots}{b + b' + b'' + \cdots} = M(a/b, a'/b', a''b'', \ldots)$." He had defined a *mean* of $c, c', c'', \ldots = M(c, c', c'', \ldots)$ as "a new quantity included between the smallest and the largest of those under consideration" in *Cours d'analyse*; see the edition of Cauchy's *Oeuvres*, series 2, vol. 3, pp. 27, 368.

cides, itself will lie between the limits $A - \varepsilon, B + \varepsilon$, and since this conclusion holds no matter how small ε may be, we can conclude that the expression (4) will be included between A and B.

Corollary. If the derived function $f'(x)$ is itself continuous between the limits $x = x_0, x = X$, this function, as it goes from one limit to the other, will vary in such a way as always to remain included between the two values A and B and to take, successively, all the intermediate values. Thus any quantity intermediate between A and B will be a value of $f'(x)$ corresponding to a value of x lying between the limits x_0 and $X = x_0 + h$, or, what amounts to the same thing, to a value of x of the form

$$x_0 + \theta h = x_0 + \theta(X - x_0),$$

where θ designates a number less than one. If we apply this remark to the expression (4), we can conclude that there exists, between the limits 0 and 1, a value of θ which satisfies the equation

$$[f(X) - f(x_0)]/(X - x_0) = f'[x_0 + \theta(X - x_0)],$$

or, what amounts to the same thing, the equation

$$(6) \quad [f(x_0 + h) - f(x_0)]/h = f'(x_0 + \theta h).$$

Since this last formula continues to hold for whatever value of x is represented by x_0, as long as the function $f(x)$ and its derivative $f'(x)$ remain continuous between the limits $x = x_0, x = x_0 + h$, we have in general, if that condition holds,

$$(7) \quad [f(x + h) - f(x)]/h = f'(x + \theta h).$$

Then, writing Δx instead of h, we have

$$(8) \quad f(x + \Delta x) - f(x) = f'(x + \theta \Delta x) \Delta x.$$

It must be noted that in equations (7) and (8) θ always signifies an unknown [nonnegative] number less than one.

Existence of Definite Integrals: *Calcul Infinitésimal*, Leçon 21 (*Oeuvres*, Series 2, Vol. 4, pp. 122–125)

Suppose that the function $y=f(x)$ is continuous with respect to the variable x between the two finite limits $x=x_0, x=X$. We designate by $x_1, x_2, \ldots, x_{n-1}$ new values of x placed between these limits and suppose that they either always increase or always decrease between the first limit and the second. We can use these values to divide the difference $X-x_0$ into elements

$$(1)\quad x_1-x_0, x_2-x_1, x_3-x_2, \ldots, X-x_{n-1},$$

which all have the same sign. Once this has been done, let us multiply each element by the value of $f(x)$ corresponding to the left-hand end point [*origine*] of that element: that is, the element x_1-x_0 will be multiplied by $f(x_0)$ [*Calcul infinitésimal* has the misprint $f(x)$], the element x_2-x_1 by $f(x_1), \ldots$ and finally the element $X-x_{n-1}$ by $f(x_{n-1})$; and let

$$(2)\quad S=(x_1-x_0)f(x_0)+(x_2-x_1)f(x_1)+\cdots + (X-x_{n-1})f(x_{n-1})$$

be the sum of the products so obtained. The quantity S clearly will depend upon

1st: the number n of elements into which we have divided the difference $X-x_0$;

2nd: the values of these elements and therefore the mode of division adopted.

It is important to observe that if the numerical values of these elements become very small and the number n very large, the mode of division will have only an insignificant effect on the value of S. This in fact can be proved as follows.

If we were to suppose all the elements of the difference $X-x_0$ reduced to a single one, which would just be that difference, we would have simply that

$$(3)\quad S=(X-x_0)f(x_0).$$

When instead we take the expressions (1) for the elements of the difference $X-x_0$, the value of S, determined in this case by equation (2), will be equal to the sum of the elements multiplied by a mean of the coefficients $f(x_0)$, $f(x_1), f(x_2), \ldots, f(x_{n-1})$.*

Moreover, since these coefficients [the $f(x_k)$] are particular values of the expression $f[x_0 + \theta(X - x_0)]$ for values of θ between zero and one, we can prove by arguments similar to those used in the seventh lecture [in proving the theorem about bounds on the differential quotient: see the appendix, p. 169] that the mean in question is another value of the same expression, corresponding to a value of θ between the same limits. We can then substitute the following for equation (2):

$$(4) \quad S = (X - x_0) f[x_0 + \theta(X - x_0)],$$

where θ will be a [nonnegative] number less than one.

To go from the mode of division we have just considered to another in which the numerical values of the elements of $X - x_0$ are still smaller, it suffices to divide each of the expressions (1) [that is, the $(x_k - x_{k-1})$] into new elements. We must then replace in the right-hand side of equation (2) the product $(x_1 - x_0)f(x_0)$ by a sum of similar products, for which sum we may substitute an expression of the form $(x_1 - x_0)f[x_0 + \theta_0(x_1 - x_0)]$, where θ_0 is a [nonnegative] number less than one; note that we will have a relation between this sum and the product $(x_1 - x_0) f(x_0)$, which is similar to the relation that exists

* Cauchy had defined a *mean* of a set of elements $\{a_1, \ldots, a_n\}$, which he designated by $M(a_1 \cdots a_n)$, to be a quantity included between the minimum and maximum of the elements of the set. At this point in the discussion of the integral, to justify the conclusion quoted above, Cauchy referred to the *Cours d'analyse*, Theorem III, Corollary (see the edition of Cauchy's *Oeuvres*, series 2, vol. 3, p. 28). This corollary, restated in modern index notation for clarity, is as follows: Suppose we are given a set of n quantities all having the same sign: $\{y_1, y_2, \ldots, y_n\}$. Consider another set $\{a_1, \ldots, a_n\}$ of n quantities, and recall that their mean $M(a_1 \cdots a_n)$ is included between the minimum and maximum of the a_k. Then the corollary states $a_1 y_1 + a_2 y_2 + \cdots + a_n y_n = (y_1 + \cdots + y_n) M(a_1 \cdots a_n)$.

Applying this to the problem in the text, let $a_k = f(x_{k-1}), k = 1, 2, \ldots, n$, and let $y_k = x_k - x_{k-1}, k = 1, 2, \ldots, n-1$, $y_n = X - x_{n-1}$. Using the corollary, Cauchy's conclusion is $f(x_0)(x_1 - x_0) + \cdots + f(x_{n-1})(X - x_{n-1}) = (X - x_0)\, M[f(x_0) \cdots f(x_{n-1})]$.

between the values of S given by equations (4) and (3). Similarly, we must substitute for the product $(x_2 - x_1) \cdot f(x_1)$ a sum of terms that can be written in the form $(x_2 - x_1)f[x_1 + \theta_1(x_2 - x_1)]$, where θ_1 again designates a [nonnegative] number less than one. Continuing in this way, we finally conclude that in the new mode of division the value of S will be of the form

$$(5) \quad S = (x_1 - x_0) f[x_0\theta_0(x_1 - x_0)] + (x_2 - x_1) f[x_1 + \theta_1(x_2 - x_1)] + \cdots + (X - x_{n-1}) f[x_{n-1} + \theta_{n-1}(X - x_{n-1})].$$

If in equation (5) we set

$$f[x_0 + \theta_0(x_1 - x_0)] = f(x_0) \pm \varepsilon_0,$$

$$f[x_1 + \theta_1(x_2 - x_1)] = f(x_1) \pm \varepsilon_1,$$

$$\vdots$$

$$f[x_{n-1} + \theta_{n-1}(X - x_{n-1})] = f(x_{n-1}) \pm \varepsilon_{n-1},$$

we can derive

$$(6) \quad S = (x_1 - x_0)[f(x_0) \pm \varepsilon_0] + (x_2 - x_1)[f(x_1) \pm \varepsilon_1] + \cdots + (X - x_{n-1})[f(x_{n-1}) \pm \varepsilon_{n-1}].$$

Then, working out these products,

$$(7) \quad S = (x_1 - x_0)f(x_0) + (x_2 - x_1)f(x_1) + \cdots + (X - x_{n-1})f(x_{n-1}) \pm \varepsilon_0(x_1 - x_0) \pm \varepsilon_1(x_2 - x_1) \pm \cdots \pm \varepsilon_{n-1}(X - x_{n-1}).$$

We may add that if the elements $x_1 - x_0, x_2 - x_1, \ldots, X - x_{n-1}$ have very small numerical values, each of the quantities $\pm \varepsilon_0, \pm \varepsilon_1, \ldots, \pm \varepsilon_{n-1}$ will be very close to zero, and therefore the same will be true for the sum $\pm \varepsilon_0(x_1 - x_0) \pm \varepsilon_1(x_2 - x_1) \pm \cdots \pm \varepsilon_{n-1}(X - x_{n-1})$, which is equivalent to the product of $X - x_0$ by a mean between these quantities [the ε_k; again, Cauchy has used the corollary about means]. Granting this, when we compare equations (2) and (7) we see that we would not change perceptibly the value of S that was calculated by a mode of division in which the elements of the difference $X - x_0$ have very small numerical values if we went to a second mode of division in which each of those elements was further subdivided into others.

Now suppose that we consider two separate modes of division of the difference $X - x_0$, in both of which the elements of the difference have very small numerical values. We can compare these two modes with a third mode, chosen so that each element, from either the first or second mode, is formed by bringing together several elements of the third mode. To satisfy this condition, it suffices for each of the values of x placed between the limits x_0 and X in the first two modes to be used in the third; and we can prove that we change the value of S very little in going from the first or the second mode to the third—and therefore, in going from the first to the second. Thus, when the elements of the difference $X - x_0$ become infinitely small, the mode of division has only an imperceptible effect on the value of S; and, if we let the numerical values of these elements decrease while their number increases, the value of S ultimately becomes, for all practical purposes [*sensiblement*], constant. Or, in other words, it ultimately reaches a certain limit that depends uniquely on the form of the function $f(x)$ and on the bounding values x_0, X of the variable x. This limit is what is called a *definite integral*.

The Fundamental Theorem of Calculus: *Calcul infinitésimal* (*Oeuvres*, Series 2, Vol. 4, pp. 151–152)

If in the definite integral $\int_{x_0}^{X} f(x)\,dx$ we let one of the two limits of integration, for instance X, vary, the integral itself will vary with that quantity. And if the limit X, now variable, is replaced by x, we obtain as a result a new function of x, which will be what is called an integral taken from the *origin* $x = x_0$. Let

$$(1)\quad \mathscr{F}(x) = \int_{x_0}^{x} f(x)\,dx$$

be that new function. We derive from formula (19) (Lecture 22) [that is, $\int_{x_0}^{X} f(x)\,dx = (X - x_0) f[x_0 + \theta \cdot (X - x_0)]$]

$$(2)\quad \mathscr{F}(x) = (x - x_0) f[x_0 + \theta(x - x_0)], \mathscr{F}(x_0) = 0,$$

θ being a [nonnegative] number less than one. Also, from formula (7) (Lecture 23) [that is, $\int_{x_0}^{X} f(x)\,dx = \int_{x_0}^{\xi} f(x) \cdot dx + \int_{\xi}^{X} f(x)\,dx$, where $x_0 \leqslant \xi \leqslant X$],

$$\int_{x_0}^{x+\alpha} f(x)\,dx - \int_{x_0}^{x} f(x)\,dx = \int_{x}^{x+\alpha} f(x)\,dx = \alpha f(x + \theta\alpha),$$

or

$$(3)\quad \mathscr{F}(x+\alpha)-\mathscr{F}(x)=\alpha f(x+\theta\alpha).$$

It follows from equations (2) and (3) that if the function $f(x)$ is finite and continuous in the neighborhood of some particular value of the variable x, the new function $\mathscr{F}(x)$ will not only be finite but also continuous in the neighborhood of that value, since an infinitely small increment of x will correspond to an infinitely small increment in $\mathscr{F}(x)$. Thus if the function $f(x)$ remains finite and continuous from $x = x_0$ to $x = X$, the same will hold for the function $\mathscr{F}(x)$. In addition, if both members of formula (3) are divided by a, we may conclude by passing to the limits that

$$(4)\quad \mathscr{F}'(x)=f(x).$$

Thus the integral (1) considered as a function of x has as its derivative the function $f(x)$ under the integral sign $\int$.

Notes

Works are cited by author and reference number.

Chapter 1

1. Of course this precept was violated often in practice; there has never been a mathematician who did not leave "obvious" steps out of proofs. But these omitted steps are supposed to be easily suppliable, at least in theory.

2. For examples of this attitude see the introduction to Bolzano's *Rein analytischer Beweis* (Bolzano [3]), Cauchy's *Cours d'analyse* (Cauchy[1]), and Abel [2] (also in Abel [1, vol. 1, p. 219]). Finally, see Poincaré [1, pp. 120–122], who stated that through the labors of nineteenth-century mathematicians "absolute rigor" had been attained.

3. Of course there are gaps of various kinds in the reasoning of Euclid and Archimedes (as there are in nineteenth-century mathematics), but these gaps were not widely noticed in the 1820s. And it was not the occasional lapses in Greek geometry that made the subject so widely admired. A look at T. L. Heath's editions of *The Thirteen Books of Euclid's Elements* or *The Works of Archimedes* will illustrate amply the claims I have made for them.

4. Cauchy's *Cours d'analyse* (Cauchy [1]; also in Cauchy [14, series 2, vol. 3, pp. ii–iii]); italics mine. Compare Cauchy's *Calcul infinitésimal* (Cauchy [15]; also in Cauchy [14, series 2, vol. 4, preface]). All page references to these two works will be from the editions reprinted in *Oeuvres complètes d'Augustin Cauchy* (Cauchy [14]).

5. These impressions come from a number of classic and modern expositions. See, for instance, Voss [1, pp. 59, 60, 95]; E. Goursat, *Cours d'analyse mathématique*, vol. 1 (Paris: Gauthier-Villars, 1910), pp. 10, 12; Ch.-J. de la Vallée-Poussin, *Cours d'analyse infinitésimale*, vol. 1 (Louvain and Paris: Gauthier-Villars, 1914), pp. 13–14; F. Klein, *Elementarmathematik vom höheren Standpunkte aus*, vol. 1 (Berlin: Springer, 1924), pp. 229–230, also pp. 166, 218, 254, and vol. 3 (Berlin: Springer, 1928) pp. 26ff, 62; Pierpont [1, especially pp. 32–35]; G. Kowalewski, *Die klassischen Probleme der Analysis des Unendlichen* (Leipzig: Teubner, 1937), pp. 38–42, 69, 104–105, 350; D. Widder, *Advanced Calculus* (New York: Prentice Hall, 1947), pp. 233–235, 242; T. M. Apostol, *Mathematical Analysis* (Reading, Massachusetts: Addison-Wesley, 1957), pp. 66, 356; J. Dieudonné, *Fondements de l'analyse moderne*

(Paris: Gauthier-Villars, 1963), pp. 48–50; R. Courant and F. John, *Introduction to Calculus and Analysis*, vol. 1 (New York: Interscience, 1965), pp. 75, 97, 511–512.

6. For examples of this reaction, see Robinson [1, chapter 10]. See also Grattan-Guinness [3, pp. 49, 60–61]. An extreme form of this reaction is in Dubbey [1]. Both Robinson and Grattan-Guinness, nevertheless, have made acute observations concerning many aspects of Cauchy's work.

7. As Cauchy remarked in the preface to the *Cours d'analyse* (Cauchy [14, series 2, vol. 3, p. ii]), some things "for those who want to make a special study of analysis" were to be found only in the notes at the back of the book.

8. A good short account of Cauchy's contribution to analysis is given by Klein [1, vol. 1, pp. 82–87]. A good brief introduction to all aspects of Cauchy's life and work, with an extensive and critically annotated bibliography, is Freudenthal's article on Cauchy (Freudenthal [1]). A faithful description of much of Cauchy's work, together with English translations of extensive selections from the original texts, may be found in Iacobacci [1]. See also Dieudonné [1, vol. 1, especially chapters 2–6].

9. *Cours d'analyse* (Cauchy [14, series 2, vol. 3, p. 19]).

10. *Encyclopédie* article entitled "Limite"; see d'Alembert and de la Chapelle [1]. The first sentence quoted is by de la Chapelle, the second by d'Alembert. D'Alembert later defined the differential quotient as the limit of the ratio of the finite differences; this definition of dy/dx was not original with d'Alembert; see Boyer [1] and Cajori [1] for numerous examples of mathematicians who defined the differential quotient in terms of limits—including Newton and Leibniz.

11. *Cours d'analyse* (Cauchy [14, series 2, vol. 3, p. 54]). Freudenthal [1] has also noted the modernity of this statement.

12. *Cours d'analyse* (Cauchy [14, series 2, vol. 3, pp. 54–57]).

13. *Cours d'analyse* (Cauchy [14, series 2, vol. 3, pp. 58–61, 121–123]).

14. The best-known pre-1860 reference is Abel's brief journal entry "Bolzano is an able man." Also, N. I. Lobachevsky read Bolzano's 1817 paper on the intermediate-value theorem; see Grattan Guinness [3, p. 52n]. We can only conjecture about what work of Bolzano's Abel had seen. Abel's journal entry was quoted by L. Sylow [1, p. 13]; Sylow added that he did not understand the reference when he first read it, believing "Bolzano" to be the name of a town, until reading about the mathematician Bolzano

in the *Enzyclopädie der Mathematischen Wissenschaften*—proof of how little Bolzano was known until historically minded mathematicians like Otto Stolz and Hermann Hankel rediscovered him. Dugac [3, p. 52; 2, p. 47] has shown recently that Weierstrass knew and used Bolzano [3] as early as 1865.

15. *Rein analytischer Beweis des Lehrsatzes dass zwischen je zwey Werthen, die ein entgegengesetztes Resultat gewaehren, wenigstens eine reele Wurzel der Gleichung liege* (Prague: 1817) (Bolzano [3]).

16. *Die drey Probleme der Rectification, der Complanation und der Cubirung, ohne Betrachtung des unendlich Kleinen, ohne die Annahmen des Archimedes, und ohne irgend eine nicht streng erweisliche Voraussetzung gelöst*, in Bolzano [4, vol. 5, pp. 67–137].

17. *Rein analytischer Beweis* (Bolzano [3, sections I–II].

18. We cite both definitions of continuity in our full discussion in chapter 4 and discuss both proofs in chapter 3.

19. *Rein analytischer Beweis* (Bolzano [3, article 6]. Bolzano gave a proof in this paper that the existence of the limit of a Cauchy sequence involves no contradiction (in effect proving the necessity of the Cauchy criterion for convergence), then used the sufficiency of the criterion to prove that a bounded set of real numbers has a least upper bound; finally, he used the least-upper-bound property to prove the intermediate-value theorem. Cauchy's own treatment of the criterion is for series; he first proved its necessity, then stated—without further argument—its sufficiency in *Cours d'analyse* (Cauchy [14, series 2, vol. 3, pp. 115–116]). He used it—albeit rarely—in discussing convergence for particular series. The Cauchy criterion and its history are treated in detail in chapter 4.

20. *Functionenlehre*, written in 1830, but not published until the twentieth century. The sections discussed are reprinted in Bolzano [4, vol. 1, pp. 144–183]; Cauchy is cited on p. 94. Compare Bolzano's *Die binomische Lehrsatz* (Bolzano [1]) discussed in Stolz [1].

21. Not mentioned in print, however, until after Weierstrass had published his own example of such a function in 1872; Boyer [1, p. 282]. For Bolzano see Kowalewski [1]. Kowalewski [1, p. 315] stated that the Bohemian Academy first aired this discovery of Bolzano's in 1921, but that Bolzano had found the function before 1834.

22. Grattan-Guinness [3, pp. 51ff]; see also Grattan-Guinness [2].

23. Compare Grabiner [1].

24. Freudenthal [2] and Sinaceur [1].

25. See chapter 4; *Cours d'analyse* (Cauchy [14, series 2, vol. 3, pp. 128–129]).

26. *Cours d'analyse* (Cauchy [14, series 2, vol. 3, p. 120]). This proof is discussed in Robinson [1, pp. 271–273], and in Grattan-Guinness [4, pp. 476–477].

27. Stolz [1, p. 265]; in 1830, *Functionenlehre* (Bolzano [2, section 75]; in Bolzano [4, vol. 1, p. 114]); compare van Rootselaar [1, p. 276].

28. Abel [2] (also in Abel [1, pp. 221–250, and footnote to theorem V, pp. 224–225]).

29. "Uniform convergence," as Grattan-Guinness [3, p. 123] has put it, "was tucked away in the word 'always,' with no reference to the variable at all."

30. For a fuller discussion of Cauchy's assumptions about completeness, see chapter 4.

31. *Cours d'analyse* (Cauchy [14, series 2, vol. 3, p. 19]) states that a real number is the limit of a sequence of rationals. Cauchy did define the product of a rational number A by an irrational number B in *Cours d'analyse* (Cauchy, [14, series 2, vol. 3, p. 337]): If there is a sequence of rationals $b, b', b'', \ldots$ [expressed verbally by Cauchy] that approach B closer and closer, then the product AB will be the limit approached by $Ab, Ab', Ab'', \ldots$. But nothing is said here about *defining* B or about the sequences $b, b', b'', \ldots,$ $Ab, Ab', Ab'', \ldots$ satisfying the Cauchy criterion. In any case, the point had already been made in antiquity; see, for example, Archimedes, *On the Measurement of the Circle*. In 1817, by contrast, Bolzano did show that the Cauchy criterion implied that a bounded sequence of real numbers had a least upper bound, though Bolzano did not yet have his own theory of real numbers fully worked out.

32. For Weierstrass see Heine [1], published in 1872. For Méray see Méray [1]. Compare Dedekind [1], published in 1872. The year 1872 was evidently a good one for real numbers; for Cantor, see G. Cantor [1], and Dauben [1].

33. Abel's writings abound with statements about rigor, especially Cauchy's rigor; see, for example, Abel [2], where the *Cours d'analyse* is termed "an excellent work, one which should be read by all analysts who love mathematical rigor." This is often quoted, for example by Freudenthal [1, p. 135].

34. For Abel's life see the article by Sylow cited in note 14, as well as the other articles in the memorial volume in which Sylow's

article appeared. Abel's life has been engagingly presented by Ore [1].

35. For Cauchy's influence on Bolzano, see, for example, Bolzano [2] (also in Bolzano [4, vol. 1, p. 94]). See also Kolman [1, part V]. Paul Funk [1, p. 133] quotes Bolzano as saying that Cauchy "is among all living mathematicians the one whom I most esteem and for whom I feel the most affinity, whose genius for discovery is responsible for some of the most important demonstrations." For Cauchy's influence on Dirichlet, see, for example, P. G. L. Dirichlet, "Sur la convergence des séries trigonometriques, . . . ," *Crelles Journal* 4 (1829): 157–169 (also in Dirichlet [1, pp. 117–139, especially pp. 118–119]). Compare Dirichlet's "Sur les integrales Euleriennes," *Crelles Journal* 17 (1837): 56–67 (also in Dirichlet [1, pp. 271–278, especially pp. 274, 277]). To document the influence on Riemann, see Riemann, "Ueber de Darstellbarkeit . . . " (Riemann [1, for example pp. 234, 238–239]) for references to Cauchy and to work of Dirichlet closely related to Cauchy's. For Abel's influence on Weierstrass see Weierstrass [1, vol. 1], where several papers study Abelian functions; and compare Biermann [1]. Weierstrass read Cauchy's works, though not until 1842, according to G. G. Mittag-Leffler [1]. Weierstrass occasionally referred directly to Cauchy; see, for example, Weierstrass [1, vol. 2, p. 90].

36. Besides the books of 1821–1823, Cauchy's rigor could be learned from numerous articles; from *Leçons sur le calcul différentiel* (1829) (Cauchy [4]), or *Leçons sur les applications du calcul infinitésimal à la géometrie* (1826–1828) (Cauchy [5]). Among the important translations, there was a German translation of *Cours d'analyse* by C. L. B. Huzler, *Lehrbuch der algebraischen Analysis* (Cauchy [6]), and a German translation of *Leçons sur le calcul différentiel* by C. H. Schnuse, *Vorlesungen über die Differenzialrechnung* (Cauchy [21]); there was also a Cauchy-based *Dei metodi analitici* (Cauchy [2]).

37. F. N. M. Moigno wrote a Cauchy-based textbook, *Leçons de calcul différentiel et de calcul intégral, redigées d'après les méthodes et les ouvrages . . . de M. A.-L. Cauchy* (Moigno [1]); a German translation was *Vorlesungen über die Integralrechnung, vorzuglich nach den Methoden von A. L. Cauchy bearbeitet* (Moigno [2]). See also C. L. Schnuse's German translation, Cauchy [20], of Cauchy's *Leçons sur les applications du calcul infinitésimal à la geometrie* (Cauchy [5]). Of importance in spreading Cauchy's work in Sweden was a book, which I have not seen, based on the *Calcul infinitésimal*, C. J. Malmsten's *Vorlesungen uber Differential- und Integralrechnung*, according to Mittag-Leffler [1]. What is essentially Moigno's version of

Cauchy's fundamental theorem of calculus may be found in Birkhoff [1, pp. 8–11].

38. For the historical comments of the mathematicians, see first Klein [1, pp. 82–87]; and then Stolz [1, especially p. 255]; Staeckel [1]; Brill and Noether [1]; and Merz [1, vol. 2, chapter 13] (though Merz was not a mathematician, he had wide competence).

39. See Merz [1, chapter 1, "The Scientific Spirit in France"]. See also Ecole Polytechnique, *Livre du centenaire 1794–1894*, vol. 1, "L'école et la science" (Paris: Gauthier-Villars, 1894). Valson [1, p. 66] states that Cauchy's lectures were heard by not only French students but faculty from France and elsewhere; Cauchy's auditors included Ampère, Sturm, Coriolis, Lamé, Dirichlet, Vallejo, Ostrogradsky, and Boniakovski. Boniakovski, for instance, expressed his pride, as late as 1882, in being a disciple of Cauchy's; see Dugac [1, p. 30].

40. See, for example, Kolman [1] and Winter [1].

Chapter 2

1. This is largely true of Newton's calculus, too, though Newton's notation was somewhat less automatic in its application; he used $\dot{x}$ for the fluxion (differential quotient) of x and $\acute{x}$ or $\boxed{x}$ for its fluent (integral). (See discussion of Newton later in this chapter.) Leibniz's notation, of course, was the one generally used in the eighteenth century on the Continent.

2. This opinion has recently been disputed by Iushkevich [4, pp. 151ff]. I do not claim that no one in the eighteenth century was interested in foundations; my thesis is that foundations were not considered of *central* importance by any mathematician until Lagrange.

3. An extensive survey is given by Boyer [1, chapter 5, "Newton and Leibniz," and chapter 6, "The Period of Indecision"]. Examples of the actual statements of Newton, Leibniz, L'Hôpital, Johann Bernoulli, Jakob Bernoulli, Bishop Berkeley, Maclaurin, d'Alembert, Euler, John Landen, and Lagrange, may be consulted in Struik [1, chapter 5].

4. An unusually thorough account in a short space of the problems studied in eighteenth-century mathematics may be found in Hofmann [1, vol. 3, chapter 8]. An extensive account may be found in M. Cantor [1, vols. 3–4] and Kline [1]. A good selection of examples of eighteenth-century mathematics may be found in Struik [1].

5. Besides the references already given, see Truesdell [1]; E. L. Ince, *Ordinary Differential Equations* (New York: Dover, 1944), appendix A, "Historical Note on Formal Methods of Integra-

tion," pp. 529–539; I. Todhunter, *A History of the Mathematical Theory of Probability* (New York: Stechert, 1931); L. E. Dickson, *History of the Theory of Numbers*, 3 vols. (Washington, DC: Carnegie Institution, 1919–1923); Goldstine [1]. Primary sources which collect a large number of results include Maclaurin [1], Euler [3, 4, 5, 8], and Lacroix [1].

6. The 1749 installment of d'Alembert's second paper on integration, "Recherches du calcul intégral," *Mémoires ... de Berlin* (1749), may appear to be an exception to our generalization, since it contains his pioneering attempt to prove the fundamental theorem of algebra. For a criticism of the inadequacies of d'Alembert's proof from the nineteenth-century point of view, however, see Gauss, "Demonstratio nova theorematis omnem functionem algebraicam ...," section 6, in Gauss [2, vol. 3, pp. 3–56, especially pp. 6, 9]. A complete translation of Gauss's paper into German may be found in E. Netto, *Die 4 Gauss'che Beweisen*, Ostwalds Klassiker 14 (Leipzig: Engelmann, 1890).

7. Proofs in number theory, though relying on formulas and making no explicit use of mathematical induction, are nevertheless more rigorous than proofs in other parts of eighteenth-century mathematics. This is in part because the problems dealt with usually are finite, and also because number theory was treated in Euclid's *Elements*, so that a rigorous model existed. For examples of eighteenth-century number theory, see Struik [1, especially pp. 36–40] for Euler's *tour de force* on Fermat's last theorem for $n = 3$ and 4.

8. Gruson [1, 2].

9. For a discussion of Lagrange's interest in foundations and his promotion of the question in the Berlin Academy, as well as for the basic doctrines of Gruson and Lagrange, see later in this chapter. Although Gruson claimed that his 1798 paper was independent of Lagrange's 1797 *Théorie des fonctions analytiques*, he certainly had read Lagrange's first exposition of their common doctrine, Lagrange [15] (also in Lagrange [9, vol. 3, pp. 439–476].

10. There were no journals devoted solely to mathematics until the nineteenth century. Other leading eighteenth-century journals, besides Berlin's, that published the mathematical contributions of leading scientists include the *Philosophical Transactions* of the Royal Society of London, the *Acta* and *Commentarii* of the Imperial Academy at St. Petersburg, the *Mémoires* of the Turin Academy, and the *Mémoires* of the Paris Academy. Mathematical contributions were also made in the form of books and pamphlets.

11. As C. Carathéodory has remarked, "We should not . . . speak of lack of rigor when we deal with methods of previous mathematicians that we can handle in a satisfactory way with the aid of techniques that we have acquired through the course of time. We should only speak of lack of rigor when some result has been obtained by means of a reasoning that cannot be logically maintained" (quoted by Struik [2, p. 125] from the introduction to Euler [6, series 1, vol. 24, p. xvi]). Carathéodory's remark helps us evaluate eighteenth-century work more fairly, but it does not change the fact the period's mathematicians *themselves* did not yet have the techniques or the interest to put their methods on a basis rigorous by modern standards.

12. For the detailed doctrines involved and for their influence on Cauchy, see first chapter 4. There are several extensive accounts of the debate; Grattan-Guinness [3, especially p. 2n] provides a list of most of them together with one of his own. Some of the relevant primary sources are excerpted in Struik [1, pp. 351–369].

13. Thus we postpone our discussion of this debate until chapter 4, where we discuss the concept of continuity and the importance of the debate about the vibrating string in clarifying that concept.

14. Thus Jakob Bernoulli remarked [1, vol. 2, "*ΕΠΙΜΕΤΡΑ*," no. 10, p. 765] that "if equals are subtracted from equals, the results are equal" ought not to be applied to arguments involving infinitesimals. He gave no example, but we can easily supply one: Suppose, as the infinitesimalists like Johann Bernoulli and L'Hôpital postulate, $a + dx = a$. Then adding dx to each side gives $a + 2dx = a + dx$. Subtracting a from each side yields $2dx = dx$, and dividing by dx yields the result $2 = 1$.

15. Often quoted, for example by Struik [2, p. 149]. The fact that this statement could be attributed to d'Alembert, a man who for his time was quite interested in foundations, illustrates the prevailing attitude. Incidentally, Dugac, [3, pp. 6–7] though agreeing that the sentiment expressed is not inconsistent with d'Alembert's views, argues that the words "la foi vous viendra" are due instead to A. Fontaine (1705–1771).

16. For a fuller treatment of this topic, see Grabiner [2].

17. The first full-scale book to receive respectful attention from mathematicians that was wholly devoted to explaining the foundations of the calculus was Simon L'Huilier, *Exposition élémentaire des principes des calculs supérieurs* (L'Huilier [1]), published in 1787 but submitted as a prize essay to the Berlin Academy of Sciences in 1784. Its genesis is discussed later in this chapter. The second

such book in the century was Lazare Carnot's *Réflexions sur la métaphysique du calcul infinitésimal* (Carnot [1]), published in 1797, but based on the manuscript submitted to the Berlin Academy prize competition; Iushkevich [4] published the manuscript. The third was Lagrange's *Fonctions analytiques* (Lagrange [16]), 1797.

18. These developments are well summarized in Hall [1].

19. On popularized science, see the brief account in Frank Manuel, *The Age of Reason* (Ithaca: Cornell, 1951), chapter 2; compare Paul Hazard, *The European Mind, 1680–1715* (Cleveland and New York: World, 1953).

20. The Newton–Leibniz controversy was by no means limited to the question of who first had invented the calculus. A short account of the philosophical side of the controversy is given in A. Koyré, *From the Closed World to the Infinite Universe* (Baltimore: Johns Hopkins, 1957), chapter 11. Simultaneous discovery, as I have argued in the case of Bolzano and Cauchy, suggests the existence of a large body of work done by earlier thinkers; see Baron [1], of which chapter 7 has a bibliography giving background for Newton and Leibniz's contributions to the calculus. Compare Boyer [1, chapter 6], and Hell, *Philosophers at War: The Quarrel between Newton and Leibniz* (New York: Cambridge U. Press, 1980).

21. An overview of British statements on this subject may be found in Cajori [1]. For a sample of the invective against Continental methods, see B. Robins, "Concerning the Nature and Certainty of Sir Isaac Newton's Methods of Fluxions and of Prime and Ultimate Ratios," 1735, in Robins [1, vol. 2, especially p. 7]; compare the book's preface by James Wilson, especially pp. x, xxiv. Incidentally, Robins's paper was written as a reply to Bishop Berkeley's attack on the calculus.

22. See Mendelsohn [1]. See also Struik [2, chapter 7, especially pp. 117, 140].

23. For Lagrange see later in this chapter. For Cauchy see chapter 1. For Weierstrass see Klein [1, vol. 1, pp. 283ff]. For Dedekind, see his own remarks in the introduction to *Stetigkeit und irrationale Zahlen*, in R. Dedekind, *Gesammelte mathematische Werke*, vol. 3, (Braunschweig: Vieweg, 1932), pp. 314–334; translated in the Dover reprint of his *Continuity and Irrational Numbers*, in R. Dedekind, *Essays on the Theory of Numbers* (New York: Dover, 1963), p. 1. For the role of schools, see also Merz [1, chapters 1–3].

24. One sign of the new attitude Lagrange brought to the subject was his claim that his "principles" of the subject were "in-

dependent of all metaphysics." See later in this chapter for further discussion.

25. In a letter to d'Alembert of 24 February 1772 (Lagrange [9, vol. 13, p. 229]). It cannot be a coincidence that Lagrange made this statement in the same year that he published the first version of his new foundations of the calculus.

26. Struik [2, p. 137]. Struik holds that this pessimism was partly due to falsely identifying the success of mathematics with that of mathematical physics; though he is correct, this does not constitute a complete explanation.

27. D. Diderot [1, section 4, pp. 180–181].

28. Among mathematicians the principal critic was Michel Rolle. Among Continental theologians, an attack on infinitesimals was made by Bernhard Nieuwentijdt. These attacks provoked replies from Leibniz, who abandoned infinitesimals in favor of a version of the limit concept. See Boyer [1, pp. 213ff, 241–242]. Compare G. Vivanti [1].

29. *The Analyst, or a Discourse Addressed to an Infidel Mathematician* (Berkeley [2]). Berkeley's attack on the ideas of the calculus can be seen, in part, as coming from his "idealist" philosophy and thus as an aspect of his criticism of the prevailing Lockeian and Newtonian ideas, including the Newtonian concepts of absolute space and absolute motion. See, for instance, Berkeley's *Principles of Human Knowledge*, sections 101–133. Berkeley held that "to be is to be perceived"; what could not be perceived, or even imagined to be an object of perception (such as absolute space or infinitesimals), did not exist; see Berkeley [2, section 5, queries 7–8]. But according to both Berkeley's explicit statements and the general emphasis of his attack on the calculus, the *Analyst* was principally motivated by his religious concerns. He replied to critics of the *Analyst* in a work even more explicitly religious in its inspiration, if not its content: *A Defence of Freethinking in Mathematics* (Berkeley [1]).

30. For a good short introduction to the thought of this period, see C. Brinton, *The Shaping of Modern Thought* (Englewood Cliffs: Prentice-Hall, 1963), chapter 4.

31. Berkeley [2]. Besides the various collections of Berkeley's works, there are other places to consult at least selections from the *Analyst*: J. R. Newman, ed., *The World of Mathematics*, vol. 1, pp. 286–293; Struik [1, pp. 333–338]; D. E. Smith, *Source Book in Mathematics*, vol. 2 (New York: Dover, 1959), pp. 627–634. The *Analyst* is worth reading in full; Berkeley is a master polemicist. I

shall discuss the mathematical particulars of Berkeley's attack later in this chapter.

32. For Maclaurin see the preface to Maclaurin [2, book I]. For d'Alembert see d'Alembert [2, 4]. In neither work did d'Alembert cite Berkeley, but the similarity of language and argument is too close for coincidence. For Lazare Carnot, see his remarks on "compensation of errors" in Carnot [1]. For Lagrange see the last sections of this chapter, p. 38ff.

33. See the first section of chapter 3.

34. The linguistic distinction between analysis and synthesis given here goes back to the Greeks and was common in the eighteenth century. Euler used the words *Analytik* and *Algebra* interchangeably; see Euler [8] and C. Boyer, "Analysis: Notes on the Evolution of a Subject and a Name," *Mathematics Teacher* 47(1954):450–462.

35. On this point see H. Butterfield, *The Origins of Modern Science, 1300–1800* (New York; Macmillan, 1958) and Hall [1].

36. In the light of Thomas Kuhn's thesis about the total breaks brought about by scientific revolutions, it may be of value to point out that this is the way mathematics usually has progressed. As Hermann Hankel said in *Die Entwicklung der Mathematik in den letzen Jahrhunderten* (Tübingen: 1884, p. 25), "In most sciences one generation tears down what another had built, and what one has established another undoes. In mathematics alone each generation builds a new story to the old structure" (quoted by Boyer [2, p. 598]). Most mathematical activity, including many major innovations, is what Kuhn has called "articulation of the paradigm"; see Kuhn [1, chapter 3, especially pp. 32–33]. I would argue, however, that at least the revolutions associated with the names of Cantor, Cauchy, and whoever first axiomatized geometry were revolutions in Kuhn's sense; see Grabiner [3]. I shall return to this question in the conclusion to this book.

37. The theory, usually attributed to Eudoxus, can be found in Euclid's *Elements*, books V and X. For references to it, see Newton [4, p. 343]; Leibniz [2, pp. 287–289], a letter to L'Hôpital, 14 June 1695; excerpt quoted by Robinson [1, p. 265]; Maclaurin [2, book 1]; L'Huilier [1; 2, *passim*].

38. Euclid, *Elements*, book V, definition 4.

39. E. J. Dijksterhuis, *Archimedes* (Groningen: 1938).

40. *On the Measurement of the Circle*, in T. L. Heath [2, pp. 91–93].

41. The details of how one shows this can be found in Euclid, *Elements*, book XII, proposition 2.

42. See the papers of Newton and Leibniz cited in note 47.

43. Most notably by Maclaurin and L'Huilier, for many specific results, in Maclaurin [2] and L'Huilier [1].

44. See *The Method* of Archimedes, discovered by Heiberg in 1906, reprinted in Heath [2].

45. See note 3 for references to fuller discussions.

46. I shall not treat here the related problem of the convergence of series. But when the question was how the successive partial sums approach the sum of an infinite series, eighteenth-century discussions are similar to those for the differential quotient of x^2. A more detailed discussion of eighteenth-century ideas on convergence will be found in chapter 4.

47. Newton [5, p. 141]. For Leibniz see first Leibniz [4]. His various views on infinitesimals may be found in his letters, published in Leibniz [2]; for example, see his letters to Varignon, 1702, in Leibniz [2, vol. 4, pp. 91–95, 106–110]. For L'Hôpital see L'Hôpital [1]. For Bernoulli see *Die Differentialrechnung* (Leipzig: Akademische Verlags-Gesellschaft, 1924 [sic]), but his views were known through L'Hôpital. For Newton's argument against infinitesimals see Newton [1, p. 39]. Both Berkeley (Berkeley [2, sections IX, XVIII]) and Lagrange, *Fonctions analytiques* (Lagrange [16, p. 3]; also in Lagrange [9, vol. 9, p. 17]) viewed neglecting infinitesimal quantities as introducing errors. For Lagrange's criticisms see later in this chapter. For a modern rigorous treatment of infinitesimals based on powerful logical methods, see Robinson [1].

48. Newton [3, pp. 49, 52]. Newton wrote o where we have h. This o notation, which was borrowed from James Gregory, was apparently intended to suggest a small increment.

49. Berkeley [2, sections XXX, XXXI]; d'Alembert [2]; Lagrange, *Fonctions analytiques* (Lagrange [16, p. 4]; also in Lagrange [9, vol. 9, p. 17]). In fairness to Newton, it should be emphasized that probably he had meant here to give an explanation, not a rigorous justification, of the neglect of h.

50. Newton [1, pp. 38–39]; Newton [5, p. 142]. Excerpts from Newton's writings on the calculus may be consulted in Struik [1, pp. 291–312].

51. Maclaurin [2, book I, p. 289], d'Alembert [1, 4], d'Alembert and de la Chapelle [1], Lacroix, [1, vol. 1, p. 6]. I shall discuss the limit concept in the eighteenth century at greater length in chapter 4.

52. Berkeley [2, sections XIII–XVI, XVIII, XXXI].

53. Berkeley [2, sections XXI–XXII]. The demonstration rests on Apollonius, *Conics*, book I, proposition 33.

54. Carnot [1, sections X, XXXVI, *et passim*]. The circumstances under which arguments like Carnot's are valid are discussed in Grattan-Guinness [1]. On Carnot see Gillispie [1], especially the essay on Carnot [1] by A. P. Iushkevich (Iushkevich [4]). Note that the 1813 revision of Carnot's work was much enlarged and Carnot's original doctrine mixed with those of others.

55. Lagrange [8] (also in Lagrange [9, vol. 7, pp. 597–599]). Compare *Fonctions analytiques* (Lagrange [16, pp. 2–3]; also in Lagrange [9, vol. 9, p. 7]).

56. Maclaurin [2, book I]. For the algebraic-style arguments modeled on the method of exhaustion, see Maclaurin [2, book II, p. 581ff]. My example is simpler than Maclaurin's usually are; I have not done justice to his ingenuity, although the principle is the same.

57. Lagrange gave no explicit reason for this equivalence. He may have argued that, if $(a - b)$ is less than any given quantity, then the Newton–d'Alembert limit concept allows the conclusion that $a = b$, so that $a - b = 0$; yet, for instance, $\lim[(a - b)/(a - b)]$ has a finite value. For Euler see Euler [3, section 83ff]; a selection may be found in Struik [1, pp. 384–386]. For Laplace see Laplace [1]; on p. 74 he stated as the first principle of the differential calculus that "two quantities whose difference can be proved to be less than any given magnitude are evidently equal to each other." For Lagrange's comment, see *Fonctions analytiques* (Lagrange [16, pp. 2–3]; also in Lagrange [9, vol. 9, p. 16]).

58. Landen [1], but Landen did not identify the power series he got with Taylor series; Arbogast [1, 2]; Gruson [1]; Lagrange [15]; and compare *Fonctions analytiques* (Lagrange [16]).

59. Cauchy [16] (also in Cauchy [14, series 2, vol. 2, pp. 276–282]). Compare *Calcul infinitésimal* (Cauchy [15, leçon 38]; also in Cauchy [14, series 2, vol. 4, p. 230]).

60. Detailed examples are given in chapters 4–5. For Lagrange's influence on Cauchy, see these chapters. Also see later in this chapter.

61. For more detail on this point, see Grabiner [2].

62. Lagrange [9, vol. 14, p. 173]. This was Lagrange's first teaching job; he was eighteen. P. E. B. Jourdain has claimed that Lagrange's views at this time were those later expounded in the *Fonctions analytiques* (Lagrange [16]). This is probably not the case;

Lagrange accepted the views of Newton and d'Alembert as late as 1760; see note 66. For Jourdain see Jourdain [1].

63. Gerdil [1]. The interest in mathematical infinity on the part of theologians was already evident in medieval times. In the eighteenth century, besides Gerdil there were Nieuwentijdt and Berkeley. For Nieuwentijdt see Boyer [1, pp. 213–214]; for the views of Fontenelle, see Boyer [1, p. 242]. In the nineteenth century, the most notable examples are Bolzano and Cantor.

64. Gerdil [1, p. 17ff].

65. This was of course Berkeley's view of 1734, though Lagrange did not say so. See note 68. Lagrange's note (Lagrange [8]) is reprinted, but without Gerdil's paper, in Lagrange's *Oeuvres* under the title "Note sur la métaphysique du calcul infinitésimal" (Lagrange [9, vol. 7, pp. 597–599]).

66. Lagrange [8, p. 18] (also in Lagrange [9, vol. 7, pp. 598–599]). Lagrange's discussion closely resembled those given by d'Alembert in his *Encyclopédie* articles and in d'Alembert [4]. D'Alembert [1] was cited by Gerdil [1].

67. Lagrange [15] (also in Lagrange [9, vol. 3, pp. 439–476]).

68. Lagrange certainly was acquainted with some of Berkeley's arguments. Unfortunately, there is no evidence about when, if ever, he read the *Analyst* (Berkeley [2]). He knew the idea of compensation of errors; Gerdil [1, p. 17]. He might have learned some of Berkeley's criticisms through their adaptation in d'Alembert's articles (note 66). We know Lagrange read the work of his friend and correspondent d'Alembert; in 1797–1799 he expressed himself in exactly the language d'Alembert had used; see notes 89–90. Lagrange did read English, as his citations of Colin Maclaurin and John Landen show, so he might well have read Berkeley [2].

69. See, for instance, his major papers on algebra, Lagrange [13] and Lagrange [12]. I shall go into this point more thoroughly in chapter 3. For the generality of his mechanics, see Lagrange [6, 7].

70. Lagrange [15] (also in Lagrange [9, vol. 3, pp. 439–476]). The purpose of Lagrange [15] basically was to give an operational calculus for the operators d and d/dx, including results like

$$e^{h(du/dx)} - 1 = h\,du/dx + h^2/2!\,d^2u/dx^2 + h^3/3!\,d^3u/dx^3 + \cdots.$$

For an account of this paper, see Koppelman [1, especially pp. 158–160].

71. The origin of his idea (Lagrange [15]; also in Lagrange [9, vol. 3, pp. 442, 445]) was Euler's *Introductio* (Euler [5]) which derived

the infinite-series expansions for all the common transcendental functions without appearing to appeal to the concept of differential quotient. See Struik [1, pp. 345–351] for a sample of Euler's style. Euler's view will be discussed in chapter 3. In the *Fonctions analytiques* (Lagrange [16]) Lagrange deduces the well-known properties of the derivative from his definition. Lagrange's notation $p, p', p'', \ldots$ is *not* our notation for derivatives, but the notation he introduced for the functions $u, u', \ldots$ in the Taylor series *is*—and Lagrange was its originator.

72. Lagrange [15] (also in Lagrange [9, vol. 3, p. 443]).

73. For instance, in 1746 d'Alembert, who previously had been little known, gained much renown and began a brilliant career by winning the Berlin prize competition on finding the causes of the winds. Both Euler and Lagrange had won the Paris Academy's prize several times. On prizes see Hankins [1].

74. The prize competition was set by the "Classe de mathématiques," which in 1784 included Lagrange, Johann (II) Bernoulli, and Johann Karl Gottlieb Schulze. Lagrange was the leading light; his long concern with the problem strongly indicates that posing it was his idea. This conclusion is supported by Hofmann [1, vol. 3, p. 68] and Iushkevich [4, p. 155].

75. Apparently because the Archimedean axiom is violated. For instance, d'Alembert said, "Given any magnitude, we can always find a magnitude larger than it." This version of the Archimedean axiom in the article "Limite" in the *Encyclopédie* (d'Alembert and de la Chapelle [1]) is clearly violated by "infinite magnitude." According to Gerdil [1, p. 2n], Gerdil, Jacquier, and Boscovitch all held "infinite magnitude" to be contradictory.

76. This phrase has a clear Lagrangian ring. Compare the full title of his *Fonctions analytiques* (Lagrange [16]) and the language of Lagrange [15].

77. "Substituted for the infinite" recalls Lagrange's comments on Gerdil's paper.

78. This requirement seems to reject the method of exhaustion in advance.

79. The whole proposal may be found in Lagrange [10, pp. 12–13]. It is reproduced in Dugac [3, p. 12].

80. Lagrange [11, p. 8]. All subsequent quotations from the committee's report are from this page; the report is quoted from extensively in Dugac [3, p. 12].

81. L'Huilier's essay (L'Huilier [1]), which used a combination of the Euclidean theory of irrationals and the limit concept, was fairly rigorous in appearance. There was some confusion, however, about what happened when—or if—a variable reached its limit. For instance, he wrote, "If a variable quantity at all stages enjoys a certain property, its limit will enjoy this same property" (L'Huilier [1, p. 167]), a statement called in the eighteenth century "the law of continuity." L'Huilier's essay was long and tedious; it was not algebraic enough to have satisfied Lagrange, and not enough of the received results of the calculus were deduced. My conclusions are based on my reading of L'Huilier's essay, since Lagrange did not go beyond the one-page critique of Lagrange [11]. L'Huilier's work was expanded into a full-scale Latin treatise, L'Huilier [2].

82. As maintained by G. Vivanti [1, p. 645]. Vivanti is certainly correct in stating that the episode increased Lagrange's dissatisfaction. Vivanti held, however, that it was Lagrange's dissatisfaction with the *winning* essay that was decisive; I believe that it was the dissatisfaction caused by reading scores of inadequate papers on a subject they claimed to understand, but clearly did not.

83. See "Advertissement" in the 2nd ed. of *Mécanique analytique* (Lagrange [7]; also in Lagrange [9, vol. 11, p. xiv]). Since this statement did not appear in the 1st ed. (Lagrange [6]), its appearance after the *Fonctions analytiques* (Lagrange [16] may be viewed as Lagrange's expression of confidence in his achievement in *Fonctions analytiques* or (I think less plausibly) as evidence that he did not worry about foundations in 1788.

84. Delambre [1, p. xxxix]. Delambre relates that this "répos philosophique" lasted until the Revolutionary government assigned Lagrange to serve on the Weights and Measures Commission.

85. *Fonctions analytiques* (Lagrange [16, p. 5]; also in Lagrange [9, vol. 9, p. 19]).

86. Lagrange [2].

87. *Fonctions analytiques* (Lagrange [16, pp. 2–3]; also in Lagrange [9, vol. 9, p. 16]).

88. *Fonctions analytiques* (Lagrange [16, p. 5], [9, vol. 9, p. 17]). Berkeley showed that the errors were compensated in the case of finding the tangent to a parabola. Incidentally, Lazare Carnot took Lagrange's statement that it was difficult to demonstrate that the errors were always compensated as praise for himself for having accomplished this difficult task! But Lagrange, since he

did not adopt Carnot's views, clearly meant it as criticism. See Carnot [2, pp. 47–48].

89. *Fonctions analytiques* (Lagrange [16, p. 4]; also in Lagrange [9, vol. 9, p. 17]). This argument is borrowed almost verbatim from the *Encyclopédie* article "Fluxion" by d'Alembert (d'Alembert [2]), though Lagrange did not cite that article. It is instructive to compare Lagrange's criticisms with Lazare Carnot's whole-hearted acceptance of all the earlier foundations and his claim that they were equivalent to his own; here Carnot said that Lagrange accepted fluxions because Lagrange said that everyone has an idea of velocity; Carnot [2, p. 167].

90. Lagrange, *Fonctions analytiques* (Lagrange [16, p. 4]; also in Lagrange [9, vol. 9, p. 17]). This argument, too, including the phrase "foreign idea," comes from d'Alembert [2]. Lagrange's attack on using the idea of motion in analysis was repeated by Bolzano, *Rein analytischer Beweis* (Bolzano [3, section II]). Bolzano's very title reinforces this point.

91. Lagrange *Fonctions analytiques* (Lagrange [16, pp. 3–4]; also in Lagrange [9, vol. 9, pp. 16–18]). Compare Berkeley [2, section IV, query 31].

92. Lagrange [2] (also in Lagrange [9, vol. 7, p. 325]). The subtangent and subsecant are, respectively, the x intercepts of the tangent and secant. The argument itself had appeared earlier in the same words in the introduction to Gruson [1] and reminds one of Berkeley's remark that the subsecant cannot become a subtangent. Lacroix, following L'Huilier, effectively met this objection, which he did not explicitly mention, by abandoning the "never surpass" requirement in 1810. See chapter 4, pp. 84–85. See also Dugac [3, pp. 13–16].

93. Lagrange [2, pp. 325–326]. He gave no reference, but this example is linked with the limit concept in d'Alembert and de la Chapelle [1] and in Newton [1], among other places.

94. This argument too was first found in Gruson [1]; it is exactly the kind of argument that would appeal to Lagrange.

95. The *Fonctions analytiques* of 1797 (Lagrange [16]) appeared in a second edition in 1813 (Lagrange [17]). Another version of the first part, omitting the applications to geometry and mechanics, appeared in 1799 as *Leçons sur le calcul des fonctions*; this work appeared in a second edition in 1806 (Lagrange [4]). Lagrange's views were used in England as the basis for the reform of the teaching of the calculus at Cambridge; his ideas were set forth as a series of notes to the English translation of Lacroix's shorter book on calculus (Lacroix [3]): *Elementary Treatise on the Differential and*

Integral Calculus, tr. Charles Babbage, John F. W. Herschel, and George Peacock (Cambridge: 1816). The *Fonctions analytiques* itself was translated into German by J.-P. Gruson. Many other works were written using Lagrange's principles; a list of these, together with other examples of Lagrange's influence, is given in Dickstein [1]. Finally, Lagrange taught these doctrines over a period of many years at the Ecole Polytechnique, thus influencing many future mathematicians.

96. Italics mine; *Rein analytischer Beweis* (Bolzano [3]). Compare sections I–II of that work, and note the references to Lagrange. Compare, further, chapters 3 and 4 of the present work.

97. Valson [1, p. 27]; the other three were Vergil, Laplace's *Mécanique celeste*, and Thomas à Kempis's *Imitation of Christ*.

Chapter 3

1. The *universal-arithmetic* view was originated by Isaac Newton, whose lectures on algebra at Cambridge (1673–1683) were published in 1707 under the title *Arithmetica universalis* (English translation 1728: Newton [6]). General symbolic notation had been introduced in 1591 by Francois Viète; Struik [1, pp. 74–81] has a selection from Viète's influential work. For examples of the general acceptance of the universal-arithmetic view in the eighteenth century see Euler [3] (reprint, p. 42); d'Alembert, "Algèbre," in the Diderot–d'Alembert *Encyclopédie*; Maclaurin [1].

2. The prevailing view in the eighteenth century was that the basis for arithmetic was the theory of magnitudes in Euclid, *Elements*, book V: that is, Eudoxus' theory of irrational ratios. See, for example, Newton [6], [2, vol. 2, p. 7]; Euler [8] (reprint, p. 42).

3. See Newton [4, section 52]; (also in Newton [2, vol. 1, p. 22]) for an early and influential statement.

4. Newton [4, section 11] (also in Newton [2, vol. 1, p. 6]).

5. The title itself, which may be translated as *Introduction to the Analysis of the Infinite*, makes this point.

6. For a sample of Euler's methods see Struik [1, pp. 348–350]. A good brief introduction to all aspects of Euler's career may be consulted in Iushkevich [1]; the article has an extensive bibliography.

7. Euler was not the only person to have tried this. Lagrange's work in this area has already been mentioned in chapter 2; for additional remarks see later in this chapter. As Lagrange himself pointed out, John Landen's *Residual Analysis* (1758, revised 1764; Landen [1]) had the same general program, though Landen did

not pay attention to the transcendental functions. In addition, much later, L. F. A. Arbogast, stimulated by Lagrange's suggestions along these lines in 1772, wrote—but did not publish—an attempt to base the calculus on the supposed algebra of infinite series; see Arbogast's remarks on his manuscript in Arbogast [1, pp. xii–xiv]. Two copies of Arbogast's manuscript exist: one, Arbogast [2], of which the Laurentian library has kindly provided me a microfilm, is in the Bibliotheca Medicea-Laurenziana, Florence, Codex Ashburnham Appendix, sig. 1840; the other is in the Ecole Nationale des Ponts et Chaussées, Paris, MS 2089. An accurate, though incomplete account of the Florence copy of this MS is given by Zimmermann [1].

8. Euler [5] (also Euler [6, series 1, vol. 8, section 4, p. 18]). Euler had recognized and discussed more general functional relations as early as 1734; the more limited definition quoted here was strictly to delineate the subject of the *Introductio*. For "analytic expression" Euler could have written equally well "algebraic expression." In his *Algebra* (Euler [8]; reprint, p. 42) he used the words "Analytik oder Algebra" as synonyms; compare Newton [4, section 52] (also in Newton [2, vol. 1, p. 22]).

9. *Fonctions analytiques* (Lagrange [16]; also in Lagrange [9, vol. 9, pp. 22–23]).

10. *Fonctions analytiques* (Lagrange [16, chapter 1, section 7]).

11. Some specific results are discussed in chapter 5. This is the principal task of Lagrange's *Fonctions analytiques* (Lagrange [16]) and his *Leçons sur le calcul des fonctions* (Lagrange [4]).

12. Once the question of convergence has been dealt with, in fact a great deal of information can be obtained about functions by studying their Taylor-series expansions. Weierstrass later exploited this idea in his theory of functions of a complex variable, retaining Lagrange's term "analytic function" to designate, for Weierstrass, a function of a complex variable with a convergent Taylor series.

13. D'Alembert [2].

14. Euler [3] (also in Euler [6, series 1, vol. 10, p. 9]).

15. *Rein analytischer Beweis* (Bolzano [3, section I]). Bolzano refers several times in the body of the *Rein analytischer Beweis* to Lagrange's work. In a recent article Kitcher [1] has argued cogently for an Aristotelian origin for Bolzano's general philosophy of mathematics; I think he is correct, but this is not the whole story. The *specifics* of Bolzano's program seem to me Lagrangian.

16. *Rein analytischer Beweis* (Bolzano [3, section II]).

17. The importance of this more general definition was first really appreciated in the work of Dirichlet, although Dirichlet did not originate it.

18. *Equations numériques* (Lagrange [18]); also in Lagrange [9, vol. 8, p. 243]), upon introducing the prime notation for derivatives.

19. Though he had been somewhat concerned with convergence, as his work on the Lagrange remainder shows, he did not expect his general computations with series to admit any but trivial exceptions. For the Lagrange remainder, see chapter 5.

20. *Cours d'analyse* (Cauchy [14, series 2, vol. 3, p. iii]).

21. *Cours d'analyse* (Cauchy [14, series 2, vol. 3, pp. iii–v]); *Calcul infinitésimal* (Cauchy [14, series 2, vol. 4, "Avertissement," p. 10]. Compare Cauchy [7] (also in Cauchy [14, series 2, vol. 8, p. 14]).

22. *Calcul infinitésimal* (Cauchy [14, series 2, vol. 4, p. 10]). In 1822 Cauchy gave the now familiar example e^{-x^2} and $e^{-x^2} + e^{-1/x^2}$ at $x = 0$; see Cauchy [16] (also in Cauchy [14, series 2, vol. 2, pp. 277–278]). Compare *Calcul infinitésimal* (Cauchy [14, series 2, vol. 4, p. 230]). But had this counterexample been the decisive reason for Cauchy's rejection of Lagrange's views, he surely would have made more of it than he did in the *Calcul infinitésimal.*

23. The explicit statement of the commutative, associative, and distributive laws together is in the work of George Peacock in the 1840s. See Koppelman [1].

24. As Viète had pointed out in the very title of the work in which he had introduced general symbolic notation, *Introduction to the Analytic Art* (my translation from Latin). The meaning "problem solving" of the term "analytic" goes back to the Greeks.

25. This is clear when one examines any of the leading eighteenth-century algebras, for example those of Newton, Maclaurin, Clairaut, and Bezout.

26. An accessible summary is given in Condorcet [1]. The first systematic survey of the subject is Lagrange's *Equations numériques*, 1798 (2nd ed. 1808; Lagrange [18]), which is discussed later in this chapter. For specific examples, the reader may consult Euler [3], Maclaurin [1], Euler [8], D. Bernoulli [1]; also see notes 28, 30, 32.

27. This view was expressed with particular force by Euler, who said that the "true sum" of the series $1 + \frac{1}{2} + \frac{1}{4} + \frac{1}{8} + \cdots$ was 2 because when the series is carried on infinitely, the error term,

which he wrote as $(x^{\infty+1}/1-x)$, becomes precisely zero; Euler [3, section 107].

28. Lambert [1]. A brief account of the method is found in Reiff [1].

29. See later in this chapter, pp. 58–59.

30. Newton [4, sections 21–29] (also in Newton [2, vol. 1, pp. 10–12]), Compare Newton [3, pp. 7–10] (also in Newton [2, vol. 1, pp. 39–41]).

31. Newton [2, vol. 1, pp. 10, 39]. Newton did not say why differing by "less than ... a tenth part" was considered close enough, but some of his other remarks suggest that it was to make his approximation resemble the way a decimal fraction approximated a real number. See, for instance, Newton [4, rule III] (also in Newton [2, vol. 1, p. 1]).

32. The exact equation for y according to Taylor's theorem would be, since $P(y) = P[(y-a)+a] = 0$,

$$y = a - P(a)/[P'(a) + (y-a)P''(a)/2! + (y-a)^2 P'''(a)/3! + \cdots].$$

Neglecting the higher powers of the small quantity $(y-a)$ gives $y = a - P(a)/P'(a)$. Newton's works of course were written before the publication of Taylor's theorem in 1715, but he knew the result himself. Maclaurin and Lambert presented Newton's method more generally than Newton. They took the general polynomial of form $y^n + qy^{n-1} - \cdots + A = 0$, with first approximation $y = a$, and gave the second approximation $y = a + p$ according to the equation $p = (A - a^{n-1} + qa^{n-1} - \cdots)/(na^{n-2} + q(n-1)a^{n-2} + \cdots)$, which would be recognized readily by any eighteenth-century mathematician as $p = -P(a)/P'(a)$; see Maclaurin [1, p. 234] and Lambert [1, pp. 148–149]. Euler and Lagrange, unlike Maclaurin and Lambert, preferred to present Newton's approximation in the language of the calculus; see Euler [3, section 234] and Lagrange's *Equations numériques* (Lagrange [18]; see also Lagrange [9, vol. 8, pp. 258–285].

33. Newton, rather than substitute his second approximation back in the original equation, used it as a first approximation to the solution of the cubic equation for p.

34. I think that he derived it from his general requirement that errors in approximations be less than 1/10. Newton did not write the equation for p in general as I have done; his treatment was entirely verbal. My reconstruction of his supposed derivation of the inequality I have written $10ec < b^2$ is as follows: Let us forget about p^3, which is very small, and consider only $cp^2 + bp + e = 0$.

We can neglect cp^2 with respect to bp, while *not* neglecting bp with respect to e, if $cp^2/bp < (1/10)(bp/e)$. Cross-multiplying in this inequality, we obtain Newton's condition $10ec < b^2$. Newton's verbal statement is in Newton [4, section 25] (also in Newton [2, vol. 1, p. 11]) and Newton [3] (also in Newton [2, vol. 1, p. 49]).

35. Euler [3, section 230]. Because $y(a) = -b, y'(a) = na^{n-1}$.

36. Possibly he had in mind the fact that the Taylor-series expansion for $y(f-a)$ was a power series in $(f-a)$, and its terms should diminish if $(f-a)$ is less than unity. In fact $|f-a| < 1$ implies Euler's condition $a^n + b < (a+1)^n$, but the converse does not hold.

37. The second term of the Taylor-series expansion for $y(f-a)$, which for the given function y is $[(f-a)^2 \cdot n(n-1)/2]a^{n-2}$, can exceed the first term $(f-a)na^{n-1}$ unless $\frac{1}{2}(f-a)(n-1) < a$.

38. He usually chose to satisfy Newton's criterion $|f-a| < a/10$, in which case the second term of the Taylor series becomes equal to or greater than the first only for $n \geqslant 21$.

39. Nor is this discussion untypical of Euler's procedure in dealing with algebraic approximations. Another example of the avoidance of general error estimates is found in Euler's discussion of the approximation method due to Daniel Bernoulli; for an exposition, see Euler [5, sections 333–355]. Euler's derivation of Bernoulli's approximation actually produced a term that that could have been used to estimate the error (Euler [5, section 346]). But instead, Euler "tested" the method as follows. He used the method, which generated a sequence of numbers $\{a_k\}$ in which the ratio a_k/a_{k-1} got closer and closer to the root, to solve the particular equation $x^2 - 3x - 1 = 0$; he then compared the seventh approximation, a_7/a_6, with the known root of that quadratic equation (Euler [5, section 338]).

40. *Cours d'analyse* (Cauchy [14, note VIII, pp. 455–457]). For instance, in proving that $\sum_1^\infty 1/k$ converged to $\pi^2/6$, Cauchy derived a formula for the nth partial sum and showed by means of extremely intricate inequality manipulation how n could be chosen to make the sum arbitrarily close to $\pi^2/6$.

41. *Fonctions analytiques* (Lagrange [16]; also in Lagrange [9, vol. 9, p. 84]); compare *Calcul des fonctions* (Lagrange [4]; also in Lagrange [9, vol. 10, p. 94]).

42. D'Alembert [3, pp. 171–215]; though published in 1768, it probably was written earlier. Of special interest to us is section I, "Reflections on Divergent and Convergent Series," pp. 171–183. (As I shall explain in chapter 4, d'Alembert used the term *con-*

vergent to mean *terms of the series are strictly decreasing, abstraction being made of the sign.*) D'Alembert discussed the application of his results to approximations made in celestial mechanics; for instance, he referred to his own *Recherches sur différens points importans du système du monde*, 2e partie, II, p. 59ff.

43. Lagrange [13]; see note 51.

44. D'Alembert [3, p. 175].

45. He did a few other such computations. For instance, fascinated by the fact that the series for sin x was considered valid for *all* x, he computed that its general ratio was (in absolute value) $x^2/2n(n+1)$. The ratio can be made less than 1 for n sufficiently large, even if $x = k$ for some very large k; d'Alembert [3, p. 181].

46. D'Alembert [3, p. 177].

47. Note that even after he had done this, it would not prove that the sum of the infinite series in fact converges to $(1+\mu)^m$, but merely that the partial sums of the infinite series are bounded. Cauchy attempted a proof of the real binomial theorem in his *Cours d'analyse*; (Cauchy [14, series 2, vol. 3, pp. 146–147]), but it used his (false) theorem that an infinite series of continuous functions was continuous. Abel gave a rigorous proof of the binomial theorem for real and complex values of μ and m in his 1826 paper (Abel [2, pp. 221–250]).

48. D'Alembert [3, pp. 177–178]. D'Alembert made the substitution $n = w + 1$, so that in his derivation the ratio was $\mu(w - m/1 + w)$ and the error less than $A\mu(m+1)/(1-\mu)\cdot(1 + w - w\mu + \mu m)$. Other than this, I have translated his explanation exactly.

49. For instance, working out $\sqrt{2} = \frac{3}{2}\cdot(1 - 1/9)^{1/2}$, he computed A, the tenth term, to be $(1\cdot 3\cdot 5\cdots 15)/9^9\cdot 2\cdot 4\cdot 6\cdots 18)$ and stated that the remainder of the series was less than $A/(1 - 1/9)$; d'Alembert [3, p. 178].

50. Lacroix's account is slightly fuller than d'Alembert's, but it too neglects the problem of absolute value; it may be found in Lacroix [2, vol. 1, p. 8ff] and is there called the method that found "the limits of the approximation," where "limit" means "bound."

51. Lagrange [13] (also in Lagrange [18] and [9, vol. 8, p. 41ff]. He used his error estimate to show that his fractions were the closest possible such approximating fractions. The Lagrange–d'Alembert correspondence indicates that Lagrange had not read d'Alembert's paper until 1769 and therefore not until after

composing his own; letter from Lagrange to d'Alembert, 28 February 1769, in Lagrange [9, vol. 13, p. 127].

52. For a brief account of this work, see Hamburg [1].

53. As for his continued fraction method, first published in 1767–1768, see note 51.

54. As he did for Newton's method, as I shall show next. The question had been raised by d'Alembert in 1768 for the binomial series.

55. One example of this is his discussion of a method due to Fontaine. D'Alembert and Condorcet, Lagrange noted, had been content to criticize the method and say it did not work well; Lagrange, instead, found algebraically the cases for which it did and did not work. Fontaine's method, he concluded, could be applied whenever the real roots of an equation differ among themselves by numbers greater than unity; Lagrange [18, note VII]. Similarly, Lagrange discussed Daniel Bernoulli's method; Euler had found a particular case in which the method did not work (Euler [8, p. 362]). Lagrange's general formulas showed that the method broke down whenever there were multiple roots; Lagrange [18, note VI].

56. Lagrange [18, note V]; also in Lagrange [9, vol. 8, especially pp. 161–163]. He did not mention d'Alembert, but the source is evident; neither the language nor the questions were commonplaces. We know Lagrange had read d'Alembert's paper because he commented on part of its contents in a letter to Almembert of 15 July 1769; Lagrange [5, p. 140].

57. Lagrange [18, note XI] (also in Lagrange [9, vol. 8, p. 258]). Though the examples he gives are polynomials, the derivation is general. This discussion, by the way, *follows* the argument about the closeness of the successive approximations.

58. *Equations numériques* (Lagrange [18]; also in Lagrange [9, vol. 8, pp. 162–163]).

59. *Equations numériques* (Lagrange [18]; also in Lagrange [9, vol. 8, p. 163]). He then gave a simple method for finding out whether the first approximation a is greater (or less) than each of the roots; Lagrange [9, vol. 8, pp. 164–165]).

60. Lagrange gave an analogous condition for the case when some roots are complex: either the real parts of those complex roots are less than the largest real root, or are greater than the smallest real root; *Equations numériques* (Lagrange [18]; also in Lagrange [9, vol. 8, p. 164]).

61. *Equations numériques* (Lagrange [18, note V]; also in Lagrange [9, vol. 8, p. 166]).

62. See, for example, *Cours d'analyse* (Cauchy [14, series 2, vol. 3, p. 413]), where Cauchy referred specifically to Lagrange's discussion of Newton's method in *Equations numériques*, though not to the specific result we have just discussed. Compare also Cauchy [14, series 2, vol. 3, p. 389].

63. *Cours d'analyse* (Cauchy [14, series 2, vol. 3, pp. 400–407, especially pp. 403, 407]).

64. Cauchy [1, p. 412]; see Fourier [1] (also in Fourier [3, vol. 2, pp. 243–253]). His concern with the closeness of Newton's approximation was aroused by Lagrange's *Equations numériques*, as Fourier himself remarked in section 2 of his paper (Fourier [3, vol. 2, p. 244]). Fourier, however, did not completely solve the problem he had set, nor did he rigorously prove what he did do.

65. *Cours d'analyse* (Cauchy [14, series 2, vol. 3, p. 415]), though without mentioning Lagrange on this page. He did mention Lagrange elsewhere in this note; see, for example, pp. 389, 413.

66. See, for instance, Maclaurin [1, p. 230ff], in which a numerical example is given.

67. Maclaurin [1, p. 233]. This was enough accuracy for some examples, though Newton would not have been satisfied.

68. This discussion is found in Lagrange [3] (also in Lagrange [9, vol. 7, pp. 260–261]). Compare the analogous result in the *Equations numériques* (Lagrange [18, sections 2 and 6]).

69. *Equations numériques* (Lagrange [18, note IV]). Compare Condorcet [1] and Cauchy's treatment of the same problem in *Cours d'analyse* (Cauchy [14, series 2, vol. 3, p. 396ff]).

70. Lagrange did *not* explicitly repeat the approximation procedure, though he did suggest transforming the original equation by letting $x = y/m$ and then solving the new equation. This latter possibility was spelled out in a bit more detail in Condorcet [1], which explicitly referred to Lagrange's work. Either of these discussions, or Maclaurin's halving method, might have suggested repeating the procedure to Cauchy—or he might well not have needed such a suggestion.

71. *Cours d'analyse* (Cauchy [14, series 2, vol. 3, p. 43]).

72. *Cours d'analyse* (Cauchy [14, series 2, vol. 3, pp. 50–51]).

73. The remark about proof is in Lagrange [9, vol. 8, pp. 19–20]. Note I, incidentally, contains a description of the continuity of a function that will be discussed in chapter 4.

74. Cauchy proved this as a consequence of the intermediate-value theorem in *Cours d'analyse* (Cauchy [14, series 2, vol. 3, pp. 381–384]).

75. *Equations numériques* (Lagrange [18], in Lagrange [9, vol. 8, p. 134]).

76. Bolzano, *Rein analytischer Beweis* (Bolzano [3, section II]; in the preface he explicitly cited Lagrange's *Résolution des equations numériques* (Lagrange [11]).

77. Grattan-Guinness [2], [3, p. 54]. For a fuller discussion of this point, see Grabiner [1].

78. Freudenthal [1, p. 134].

79. "Note on the formulas which result from using the signs $<$ or $>$, and on the means between several quantities."

80. See, for example, Newton [6, rule II]; quoted in Struik [1, p. 96].

81. See, for example, Cauchy [17].

Chapter 4

1. Grattan-Guinness [3, pp. 76–77] also has noted that Cauchy's pre-1821 work showed few signs of the program of rigorizing analysis, though he gave this fact a different interpretation, saying that the program is borrowed from Bolzano. Recently Freudenthal [2] has put forward the same explanation I have given. As indicated in chapters 2 and 3, both Bolzano and Cauchy easily could have drawn their program from the work of Lagrange and, to some extent, from d'Alembert, Euler, and Berkeley. In fact, however, Cauchy's 1814 paper already shows more rigor than most eighteenth-century work. On the eighteenth-century origins of Cauchy's program, see also Grabiner [1].

2. See chapter 1, pp. 13–14.

3. I shall document the claim that Lagrange's work was "outstanding" in chapter 5 and explain in what area its superiority lay.

4. See chapter 2, p. 43ff.

5. Grattan-Guinness has claimed that Bolzano was Cauchy's main source. I do not expect to be able to prove that Cauchy *never* had read Bolzano's *Rein analytischer Beweis* of 1817. I can show, however, that there was nothing in Bolzano's work on convergence and continuity used by Cauchy that Cauchy could not have derived from elsewhere—indeed, from sources that Bolzano himself almost certainly knew.

6. Lacroix [1, 2].

7. Lacroix [2, vol. 1, p. xviii]. See chapter 2 on the need for textbooks for the newly expanded eighteenth-century scientific community.

8. Lacroix [1, vol. 1, preface, especially p. xxiv]. See also Lacroix [2, vol. 1, preface, pp. xviii–xix]. For examples of his use of different methods see, for instance, Lacroix [1, vol. 1, pp. 82–95] (Lagrange's power series); Lacroix [1, vol. 1, pp. 6, 189–194] (limits); Lacroix [1, vol. 1, p. 193] (differentials treated as an independent concept); Lacroix [1, vol. 1, pp. 192–193] (on the rigor of ancient geometry); Lacroix [1, vol. 1, p. 10] (on infinitesimals and their equivalence to limits). Compare Boyer's assessment of Lacroix in Boyer [1, p. 265].

9. *Cours d'analyse* (Cauchy [14, series 2, vol. 3, p. 19]). Compare *Calcul infinitésimal* (Cauchy [14, series 2, vol. 4, p. 13]).

10. Boyer [1, pp. 256, 272–273].

11. Lacroix [2, vol. 1, p. 13]; italics in original.

12. Lacroix [2, vol. 1, pp. xlix, 14].

13. Newton's *Philosophiae naturalis principia mathematica* (1687) solved the most important problems of physics. For instance, in an elliptical orbit with a central force directed toward one focus of the ellipse, the force needed to hold the planet in the orbit is inverse square. Under an inverse-square force proportional to the mass of a body, a spherical body acts as though all its mass were concentrated at its center. If a central force acts on a body, the line connecting the body to the center of force sweeps out equal areas in equal times. The *Principia* also contains Newton's laws of motion and the law of universal gravitation. Several modern editions exist. The most easily available is the edition by Cajori (Newton [1]). Compare chapter 2.

14. Newton [1, section I, lemma VII, p. 32]; italics mine.

15. Newton [1, scholium to lemma XI, p. 39]; italics mine.

16. Newton [1, scholium to lemma XI, p. 39]; italics mine.

17. D'Alembert [2].

18. See chapters 2 and 3.

19. The "error" z by which the slope of the secant differs from the slope of the tangent can thus be made—as in algebraic approximations—less than any given quantity. D'Alembert's treatment of the tangent to the parabola is clearly, though not explicitly, an answer to Berkeley's treatment of the same question, with an almost identical diagram, in Berkeley [2, sections

XXI–XXIV]. Berkeley, unlike d'Alembert, explicitly had called z an "error" because "a secant cannot be a tangent.... Be the difference ever so small, yet still there is a difference" (Berkeley [2, section XXIV]. Understanding limits in terms of inequalities—a step d'Alembert almost, but not quite, took—is the way to answer this objection. For d'Alembert's discussion, see d'Alembert [2]. Extensive selections can be found in Struik [1, pp. 342–345].

20. If z is positive, the difference $(a/2y) - (a/2y + z)$ will be less than ε if $z < \varepsilon(4y^2/a - 2y\varepsilon)$, which would have been an easy computation for d'Alembert had he cared to make it. But he did not make it. He did not understand his words in these terms or appreciate that much more than meets the eye is involved in taking z small enough to make the difference as small as desired: in particular, on what does z depend?

21. Lacroix [3, p. 7]. This one-volume work was an abridgment of his three-volume treatise, and emphasized limits more, and power series less, than the longer work. The theorem proved here, that the limit of a product is the product of the limits, is stated in d'Alembert and de la Chapelle [1] and had (according to them) appeared earlier in de la Chapelle's *Géométrie*. The analogous theorem for ratios (without special attention to the case where the denominator may be zero) was proved by L'Huilier [2, p. 18] by an argument exactly like that given for products in Lacroix [2, vol. 1, p. 15].

22. Boyer [1, pp. 256, 272–273].

23. L'Huilier [2, pp. 17–18], where L'Huilier applied this broadened definition only to alternating series. The two examples he gave were the series $1 - p + p^2 - \cdots$ and the series $1 - \frac{1}{2} + \frac{1}{3} - \cdots$. His general definitions, which applied only to increasing or decreasing sequences of values of variables, always included the restrictions (L'Huilier [2, p. 1]).

24. Lacroix [2, vol. 1, p. xlix].

25. Lacroix [2, vol. 1, p. 140]; compare Lacroix [1, vol. 1, p. 82].

26. Lacroix [2, vol. 1, p. 14].

27. Lacroix [2, vol. 1, p. 15].

28. Berkeley [2, query 31].

29. Lacroix [1, vol. 1, p. 192].

30. Maclaurin [2, book II, p. 422].

31. D'Alembert [4, vol. 5, p. 247]. Lacroix cited d'Alembert as the source for his own discussion; Lacroix [1, vol. 1, p. xxx].

32. *Calcul infinitésimal* (Cauchy [15, leçon 3]; also in Cauchy [14, series 2, vol. 4, p. 22]).

33. *Calcul infinitésimal* (Cauchy [14, series 2, vol. 4, p. 22]); italics mine, notation his.

34. See, for instance, *Fonctions analytiques* (Lagrange [9, vol. 9, pp. 59–62]) for a number of examples of functions without derivatives at particular points.

35. I have given an example of the translation of the definition in chapter 1. For an example of Cauchy's determination of the n corresponding to ε for a complicated case, see *Cours d'analyse* (Cauchy [1, note VIII]; also in Cauchy [14, series 2, vol. 3, pp. 456–458]); compare chapter 3.

36. *Cours d'analyse* (Cauchy [14, series 2, vol. 3, p. 43]).

37. *Cours d'analyse* (Cauchy [14, series 2, vol. 3, p. 44]. Cauchy did not compute the deltas and epsilons, however. He used the identity $\sin(x + a) - \sin x = 2 \sin(\frac{1}{2}a) \cos(x + \frac{1}{2}a)$, and noted that, whatever x may be, $\sin(\frac{1}{2}a)$ "decreases indefinitely" with a. Cauchy here gave no reason for this indefinite decrease, but it was well known that, for $0 < a < \pi/2$, $\sin a < a$. See, for example, *Fonctions analytiques* (Lagrange [9, vol. 9, p. 54]).

38. *Rein analytischer Beweis* (Bolzano [3, section IIa]). He understood in practice that $|f(x + w) - f(x)|$ is needed here. Cauchy and Bolzano did not seem to appreciate that they were assuming, in effect, that given an ε, their δ works for all x.

39. Arguments like this go back to Euclid, *Elements*, book I, proposition 1. See, in this connection, the remarks of Bolzano in the introduction to his *Rein analytischer Beweis* (Bolzano [3]). It is probably worth remarking also that attempts to characterize continuity in geometry go back at least to Aristotle.

40. Chapter 3. *Résolution des équations numériques* (Lagrange [1, note 2]). Recall that Bolzano criticized Lagrange's proof, quite properly, as depending on the idea of time. Bolzano [3, section IIc; Ostwalds Klassiker, p. 6].

41. *Equations numériques* (Lagrange [1, note 2]); also in Lagrange [9, vol. 8, pp. 136–139].

42. See Boyer [1, p. 256].

43. Lacroix [3, section 60, p. 82]. Compare the parallel discussions (though without mention of continuity) that go back to Newton; see earlier in this chapter.

44. There is an extensive literature on this problem. An excellent recent account is that given by Truesdell [1, especially

pp. 237–300]. The leading papers on the vibrating-string controversy are excerpted, with useful explanatory remarks, in Struik [1, pp. 351–358; see also "Note on the Emergence of the Concept of Function," p. 358], and Jourdain [2], who discusses the work of Fourier also. Grattan-Guinness [3, chapter 1] has cited all the relevant secondary literature, beginning with Riemann, in his footnotes.

45. And, of course, conversely. Reproducing d'Alembert's original notation would add nothing to the clarity of my account; it is reproduced in Struik [1, pp. 351–358].

46. *Hist. Berl.* (1750); quoted by Struik [1, p. 361].

47. See Jourdain [2] or Reiff [1] for Fourier and Dirichlet's solution to the problem of the generality of trigonometric-series representations of functions. For the way the various views changed throughout the eighteenth century, see Truesdell [1].

48. Quoted by Jourdain [2, p. 675]. *Algebraic*, a term established in this usage by Descartes, meant anything represented by a finite polynomial; *mechanical*, everything else. *Mechanical* had also been used for curves generated by machinery, which did the drawing. *Transcendental* was introduced by Leibniz for curves representable by infinite series. For *discontinuous* in this context, read Euler's "not representable by the same formula throughout," and see later in this chapter. "Produced by a voluntary movement of the hand" can mean almost anything; it could perhaps be taken to refer to any general dependence relation.

49. Euler [5, vol. 2, chapter 1] (also in Euler [6, series 1, vol. 9]).

50. *Mémoires de l'Académie des Sciences*, Paris, 1771.

51. Arbogast [3, p. 5].

52. Arbogast [3, p. 9]. Compare Lacroix [3, section 60, p. 82]: "The smaller the increments of the independent variable, the closer the successive values of the function are to each other."

53. Arbogast [3, p. 9].

54. Arbogast [3, p. 9].

55. Arbogast [3, p. 10]. Conversely, Arbogast seems implicitly to hold that a continuous function is necessarily contiguous (Arbogast [3, p. 11]). Grattan-Guinness [2, p. 53] implies that since Cauchy followed Bolzano in talking about intervals of continuity, Cauchy borrowed his definition from Bolzano. But this contention is not tenable in the face of widespread eighteenth-century usage. Furthermore, Cauchy, three years before Bolzano's 1817 paper, had an independent reason for concern with intervals of continuity—his study of the integrals of piece-

wise continuous functions; see in this chapter. It is possible that the stress on continuity over intervals helped obscure the distinction between uniform continuity and continuity at a point.

56. Lacroix [3, p. 82].

57. Bolzano [2, p. 16], though in a somewhat different connection. This reference to Lacroix is cited by Grattan-Guinness [3, p. 77, note 23].

58. Cauchy [10] (also in Cauchy [14, series 1, vol. 1, pp. 329–475, especially p. 332, and compare p. 402]). I have modernized the notation for the definite integral.

59. Jourdain [2, pp. 681–682, 688].

60. Cauchy [9] (also in Cauchy [14, series 1, vol. 1, pp. 402–403]). The function in question is an indefinite integral; one example Cauchy gives is $\arctan(1/\cos z)$, which has a jump discontinuity at $z = \pi$. Freudenthal [2, p. 380] has taken this discussion to mean that Cauchy already had the "full-fledged idea of continuity" in 1814. This claim had already been made on other grounds by Jourdain [2], who pointed out that two theorems in Cauchy's memoir (Cauchy [14, series 1, vol. 1, pp. 428, 441]) explicitly assume that a function is continuous. The proofs, however, require only "no jumps." Perhaps Cauchy did already have the full-fledged idea of continuity, but this is not demonstrated by his use of the no-jumps property in 1814.

61. This result was long known and believed; I shall return to the result and its applications in chapter 5. See *Fonctions analytiques* (Lagrange [16]; also in Lagrange [9, vol. 9, pp. 28–29]).

62. In Lagrange's theorem, hP is the function $hp + h^2q + \cdots$, but its form is irrelevant to the continuity argument presented here.

63. *Fonctions analytiques* (Lagrange [9, vol. 9, p. 28]); italics mine.

64. In modern notation, Lagrange is saying that hP continuous at $h = 0$ means that given an $\varepsilon > 0$, there exists h_0 such that $|h_0P| < \varepsilon$ and, moreover, $|hP| < \varepsilon$ for $h < h_0$. Compare the equivalent characterization in Lagrange's *Calcul des fonctions* (Lagrange [4]; also in Lagrange [9, vol. 10, p. 87]) where, however, the term *continuous* does not occur. Notice that the choice of h_0 does not depend on x.

65. For Bolzano see, for example, Bolzano [2, p. 170], *Rein analytischer Beweis* (Bolzano [3, section V]). The *Fonctions analytiques* was in his library; see Bolzano [2, K. Rychlik, "Anmerkungen" p. 23]. For Cauchy see, for example, *Calcul infinitésimal* (Cauchy [14, series 2, vol. 4, pp. 9–10]) and *Calcul différentiel* (Cauchy [14, series 2, vol. 4, p. 268]).

66. Bolzano [2, pp. 15–16].

67. Bolzano [2, p. 16]; italics his. Among the "others" who used this definition is Klügel [1, vol. 4, p. 550], who is cited by Bolzano [2, p. 15].

68. Bolzano [2, p. 14].

69. See the appendix for these proofs.

70. On Bolzano's unfortunate lack of influence, see chapter 1.

71. *Cours d'analyse* (Cauchy [14, series 2, vol. 3, p. 130; compare pp. 121, 136]).

72. Especially Euler [5, *passim*].

73. Gauss [1] (also in Gauss [2, vol. 3]). See, for example, Bell [1, p. 291], Reiff [1, p. 161].

74. *Cours d'analyse* (Cauchy [14, series 2, vol. 3, p. 114]). For the Cauchy criterion, see later in this chapter, p. 102ff.

75. See, for instance, Klügel [1, "Convergirend, annährend," section 2]. Note the title of this article. The term *converge* in this sense is also used in optics. Lagrange used *converge* in Klügel's sense; see Lagrange [13] (also in Lagrange [9, vol. 2, p. 541]), where he spoke of a series diverging after beginning to converge; the context makes clear that he meant that the terms get bigger after decreasing for a while. The same usage may be found in d'Alembert [3, especially p. 173] and in Legendre [1, vol. 3, p. 437]. Euler also used *converge* for *terms decrease*; see Euler [2] (also in Euler [6, series 1, vol. 14, pp. 586, 588]).

76. Both definitions exist together in Hutton [1]; the article "Converging Series" has the *n*th term go to zero, but the term *convergence* is defined in the article "Series" as *having a finite sum*. The same situation exists in *Dictionnaire encyclopédique des mathématiques*, "Convergent," by d'Alembert, and "Série ou suite" by Condorcet (Condorcet [2]). Nevertheless, Condorcet clearly understood what was involved, since he specifically mentioned the harmonic series; see Condorcet [2].

77. For Oresme, see H. L. Busard's edition of Nicole Oresme, *Questiones super geometriam Euclides* (Leiden: E. J. Brill, 1961), pp. 6, 76. Jakob Bernoulli's work is in "Proportiones arithmeticae de seriebus infinitis earumque summa finita" (Jakob Bernoulli [1, vol. 1, pp. 375–402]). Selections may be consulted in Struik [1, pp. 320–324]. For Euler see Euler [2] (also in Euler [6, series 1, vol. 14, pp. 87–100]). For Condorcet, see Condorcet [2].

78. *Cours d'analyse* (Cauchy [14, series 2, vol. 3, pp. iv, 114]). Some such "uncritical manipulations" may be found in Euler [2,

pp. 204–237] (also in Euler [6, series 1, vol. 14, pp. 585–617, especially pp. 591–597]).

79. Lacroix use *converge* here to mean *nth term goes to zero.*

80. This part of the sentence is equivalent to convergence in the modern sense. The "true value" is the expression that produced the series; for the binomial series, for instance, it would be $(1 + x)^m$; Lacroix [2, vol. 1, p. 5]. Compare d'Alembert [3, p. 183].

81. *Cours d'analyse* (Cauchy [14, series 2, vol. 3, p. iv]).

82. Maclaurin [2, book I, p. 289].

83. A sequence $\{u_n\}$ is now called a Cauchy sequence if for any ε, there is an N such that $|u_n - u_m| < \varepsilon$ whenever $n, m > N$.

84. *Cours d'analyse* (Cauchy [14, series 2, vol. 3, p. 114]).

85. *Cours d'analyse* (Cauchy [14, series 2, vol. 3, pp. 115–116]; italics mine.

86. *Rein analytischer Beweis* (Bolzano [3, section 7; Ostwalds Klassiker, p. 21]. The proof he gave of this fact, as Jourdain has remarked in "Anmerkungen" to the edition (Bolzano [3; Ostwalds Klassiker, p. 42]), shows only that the limit X can exist without there being a contradiction.

87. Lacroix [2, vol. 1, pp. 8–9], which is an exposition, in somewhat clearer notation, of results in d'Alembert's 1768 paper (d'Alembert [3]). See chapter 3, note 50.

88. Lacroix [2, vol. 1, p. 8]; italics mine. An additional resemblance between Lacroix's and Cauchy's characterizations of these finite expressions is that Lacroix spoke of the "diverse" approximations, Cauchy of the "diverse conditions."

89. *Cours d'analyse* (Cauchy [14, series 2, vol. 3, pp. 116–117]). Compare Lacroix–d'Alembert, chapter 3, pp. 62–63.

90. *Cours d'analyse* (Cauchy, [14, series 2, vol. 3, p. 117]).

91. Euler [1]. Eneström [1] has somewhat overenthusiastically seen this paper as a source of the general Cauchy criterion. For another possible source, this time from Gauss, see Kurdyumova [1], again just for a specific case.

92. There were a few other proofs available. Lacroix showed the divergence by interpreting the harmonic series as the logarithm of an infinite number (Lacroix [3, vol. 1, pp. 5–6]).

93. *Cours d'analyse* (Cauchy [14, series 2, vol. 3, pp. 130–131]). The sum of this series was given in 1673 by Leibniz, who used its telescoping property; Reiff [1, pp. 42–43]. Cauchy's proof also used the telescoping property. For instance, if $n + 1$ is odd, then

$$1/\overline{n+1} - 1/\overline{n+2} + \cdots \pm 1/\overline{n+m} = (1/\overline{n+1} - 1/\overline{n+2}) + (1/\overline{n+3} - 1/\overline{n+4}) + \cdots$$

lies between $1/\overline{n+1}$ and $+1/\overline{n+1} - 1/\overline{n+2}$. The series converges now by the Cauchy criterion.

94. Besides the alternating series proof, Cauchy implicitly used something very close to the Cauchy criterion in his treatment of the existence of the definite integral; see chapter 6. Had he given a correct proof of the convergence of an absolutely converging series, he would have had to use it there too; see note 106.

95. See in particular the work of d'Alembert, chapter 3.

96. *Cours d'analyse* (Cauchy [14, series 2, vol. 3, pp. 114–115]). Since the nth remainder $x^n/1 - x$ converges to 0 if $|x| < 1$ and n increases.

97. Lacroix [2, vol. 1, p. 8ff]. The table of contents explicitly cites d'Alembert's paper.

98. *Cours d'analyse* (Cauchy [14, series 2, vol. 3, p. 121]). Cauchy implicitly assumed that such a "greatest of these limits" or "limit of the greatest values" exists; we would call this object the lim sup and assert its existence from the completeness of the real numbers.

99. Nor has it been improved on much since, as can be seen from looking at the proof of the root test in a modern advanced calculus text. See, for example, D. Widder, *Advanced Calculus* (New York: Prentice-Hall, 1947), chapter 9.

100. *Cours d'analyse* (Cauchy [14, series 2, vol. 3, pp. 121–122]), assuming the comparison test. He showed analogously that if $k > 1$, then the series diverges. In his treatment of the hypergeometric series in 1813, Gauss also made careful use of the comparison with a convergent geometric progression—as indeed the very term *hypergeometric*, dating back to the seventeenth century, suggests. Though Cauchy did not need Gauss's paper for the development of the theory of convergence in the *Cours d'analyse*—indeed, he did not even mention the general hypergeometric series—it is hard to imagine that Cauchy had not read Gauss's paper. Bolzano also used the comparison test, with respect to the geometric series $\Sigma\ 1/2^k$; however, in the *Rein analytischer Beweis* he gave no general convergence tests. Comparisons with the series $\Sigma\ 1/2^k$ are common in the seventeenth and eighteenth centuries; see Newton [4] and compare Newton's source, Euclid, *Elements*, book X, proposition 1.

101. *Cours d'analyse* (Cauchy [14, series 2, vol. 3, p. 123]).

102. Leçon 38 (Cauchy [14, series 2, vol. 4, pp. 226–228]). The derivation is somewhat more sophisticated.

103. *Cours d'analyse* (Cauchy [14, series 2, vol. 3, p. 63]). The theorem just cited is a simple corollary of this theorem: If $\lim_{x\to\infty} f(x+1)/f(x) = k$, then $\lim_{x\to\infty} f(x)^{1/x} = k$. Cauchy allowed k to be infinite; see *Cours d'analyse* (Cauchy [14, series 2, vol. 3, pp. 58–61]). The proof of this theorem is intricate and ingenious. It somewhat resembles an argument in Gauss [1, section iv]. Alternatively and, I believe, more plausibly, it may have been constructed by analogy (changing sums to products) from Cauchy's own earlier proof of the theorem that if $\lim_{x\to\infty} \cdot f(x+1) - f(x) = k$, then $\lim_{x\to\infty} f(x)/x = k$; see *Cours d'analyse* (Cauchy [14, series 2, vol. 3, pp. 54–57]). The technique Cauchy used to prove this last-stated theorem is similar to, though somewhat easier than, that used in proving a theorem about derivatives in the *Calcul infinitésimal* (Cauchy [15, leçon 7]); the history of the proof technique used in leçon 7 of the *Calcul infinitésimal* is discussed in chapter 5 and in Grabiner [4]; compare also Dugac [2].

104. See, for instance, Lacroix [2, vol. 1, pp. 27–39] and Lagrange's *Calcul des fonctions* (Lagrange [4], [9, vol. 10, pp. 48–49]). For Cauchy's statement of the theorem for positive terms, see *Cours d'analyse* (Cauchy [14, series 2, vol. 3, p. 127]); for power series, see Cauchy [14, series 2, vol. 3, pp. 140–141].

105. *Cours d'analyse* (Cauchy [14, series 2, vol. 3, pp. 134–135]). Let the two series to be multiplied each be $1/1 - 1/\sqrt{2} + 1/\sqrt{3} - 1/\sqrt{4} + \cdots$. I have been unable to find any indication that this property of this series was known before; see Reiff [1, pp. 170–171]). The series itself was discussed in the eighteenth century; see, for example, Euler [7] (also in Euler [6, series 1, vol. 15, section 13, p. 83]). I am indebted to Kenneth Manning for this reference.

106. He tried to prove this by means of the comparison test, comparing Σu_k with the series $\Sigma|u_k|$. But the partial sums of a series can be bounded without the series converging: for instance, $1 - 1 + 1 - 1 + 1 - \cdots$. It is a shame no alert student noticed the error in Cauchy's proof, since Cauchy certainly would have been able to correct it, using the Cauchy criterion. See *Cours d'analyse* (Cauchy [14, series 2, vol. 3, p. 129]).

107. *Cours d'analyse* (Cauchy [14, series 2, vol. 3, p. 129]).

108. *Cours d'analyse* (Cauchy [14, series 2, vol. 3, p. 137]). Incidentally, Cauchy's notation for radius of convergence was usually $x = 1/A$; Lacroix had used $x = 1/a$ in dealing with the binomial series. This is another one of the "incidental" resemblances between the language or notation of the two men that suggests influence; see Lacroix [2, vol. 1, p. 7].

109. For a history of convergence tests from Cauchy on, see the appendix to Grattan-Guinness [3, pp. 132–151]; compare Reiff [1, pp. 196–211].

110. *Cours d'analyse*, (Cauchy [14, series 2, vol. 3, p. ii]).

111. *Cours d'analyse*, (Cauchy [14, series 2, vol. 3, pp. iv–v]).

112. *Cours d'analyse* (Cauchy [14, series 2, vol. 3, pp. 132–135]). See also chapter 4; compare chapter 3 and *Calcul infinitésimal* (Cauchy [14, series 2, vol. 4, p. 230]).

113. Cauchy used the term *imaginary* where we would use *complex*.

114. *Cours d'analyse* (Cauchy [14, series 2, vol. 3, pp. 222–225]).

115. *Cours d'analyse* (Cauchy, [14, series 2, vol. 3, pp. 230–238]).

116. *Cours d'analyse* (Cauchy [14, series 2, vol. 3, pp. 243–247]).

117. *Cours d'analyse* (Cauchy [14, series 2, vol. 3, pp. 146–147]).

118. The false theorem for the complex case is stated in *Cours d'analyse* (Cauchy [14, series 2, vol. 3, p. 234]); it is proved by reference to the theorem for the real case (Cauchy [14, series 2, vol. 3, p. 120]).

119. Grattan-Guinness [3, p. 78], Lakatos [1, p. 130].

120. *Cours d'analyse* (Cauchy [14, series 2, vol. 3, p. 146]), where he referred to it as Theorem I, sec. I., that is, Cauchy [14, series 2, vol. 3, p. 120].

121. *Cours d'analyse* (Cauchy, [14, series 2, vol. 3, p. 257]).

122. *Cours d'analyse* (Cauchy, [14, series 2, vol. 3, pp. 248–250]); compare the simpler argument for the real case in Cauchy [14, series 2, vol. 3, pp. 147–148].

123. *Fonctions analytiques* (Lagrange [9, vol. 9, pp. 45–47]).

124. *Cours d'analyse* (Cauchy [14, series 2, vol. 3, p. 251]).

125. *Cours d'analyse* (Cauchy [14, series 2, vol. 3, p. 252]).

126. *Cours d'analyse* (Cauchy [14, series 2, vol. 3, p. 253]).

127. *Cours d'analyse* (Cauchy [14, series 2, vol. 3, p. 254]).

128. I have, of course, by no means exhausted the richness of the *Cours d'analyse*. See Freudenthal [1] and chapter 1. See also the literature cited in chapter 1, note 8, and Freudenthal's bibliography, especially on complex variables.

Chapter 5

1. *Calcul infinitésimal* (Cauchy [14, series 2, vol. 4, p. 13]).

2. *Calcul infinitésimal* (Cauchy [14, series 2, vol. 4, pp. 22–23]).

3. This proof is given in translation in the appendix. The notation is Cauchy's throughout. The proof and its historical antecedents are discussed later in the chapter. Notice how the inequality translation of the definition of derivative resembles that for the definition of limit; see chapter 1 and compare *Cours d'analyse* (Cauchy [14, series 2, vol. 3, p. 54]). On this theorem and its history see Grabiner [4], and compare Dugac [1].

4. With the possible exception of Bolzano, who knew in 1816 that $\phi(x + \omega) + \phi(x) = \phi'(x) + \Omega$ where Ω can be made as small as desired when ω is small; this, however, was a *property* of the derivative for Bolzano, not a definition. The property can already be found in Lagrange's *Fonctions analytiques*, as I shall show. For Bolzano see Stolz [1, p. 264]. Bolzano's major work on derivatives, the *Functionenlehre* (Bolzano [2]), dates from the 1830s, and cites Cauchy's work; see, for example, Bolzano [2, p. 94].

5. Lagrange, *Calcul des fonctions* (Lagrange [9, vol. 10, p. 87]). Compare *Fonctions analytiques* (Lagrange [9, vol. 9, p. 77]) for a similar formulation.

6. Euler [3, section 122]. This criterion, incidentally, is also the ultimate source of Cauchy's theory of infinitesimals of different orders. The criterion was adopted by S.-F. Lacroix, from whose work Cauchy—as similarities strongly suggest—probably derived his theory. For instance, compare their statements and proofs of the theorem that the sum of infinitesimals of orders $n, n', n'', \ldots$ is a new infinitesimal of order n if $n < n' < n'' < \cdots$. See Lacroix [1, vol. 1, pp. 16–18] and Lacroix [2, vol. 1, pp. 15–18]; compare Cauchy, *Cours d'analyse* (Cauchy, [14, series 2, vol. 3, pp. 38–42, 64–65]). As indicated in chapter 4, Cauchy knew the work of Lacroix.

7. Euler [3, sections 253–254]. Similarly, since the a^2 term can be made to exceed the sum of all that follow it, Euler argued that if x is a relative maximum, then d^2y/dx^2 must be negative; and if x is a minimum, then d^2y/dx^2 must be positive. He added that similar considerations applied to higher-order examples (Euler [3, section 255]). The use of these considerations in Euler's treatment of maxima and minima was highlighted by Iushkevich [2]. Iushkevich also pointed out that there was a kinship between this work of Euler's and Lagrange's treatment of maxima and minima using the Taylor-series remainder. For details on Lagrange's theory of extrema, see later in this chapter.

8. Maclaurin [2, sections 261, 858–859].

9. Possibly mediated by a paper of Arbogast (Arbogast [2]). See Grabiner [2].

10. *Fonctions analytiques* (Lagrange [9, vol. 9, pp. 28–29]).

11. This case is strongly supported by Iushkevich [2].

12. *Fonctions analytiques* (Lagrange [9, vol. 9, p. 282]).

13. Except, possibly, at some finite number of isolated points. He thought he had proved this; *Fonctions analytiques* (Lagrange [9, vol. 9, pp. 22–23]).

14. This definition is borrowed from Euler [5]. Note that Euler used this definition because the work (Euler [5]) is solely the study of infinite analytic expressions; elsewhere he recognized and used a broader definition of function.

15. *Fonctions analytiques* (Lagrange [9, vol. 9, pp. 28–29]). Arbogast earlier (1789) had tried to prove this too, though in a very different way than Lagrange. Arbogast's method was to choose i so that each term of the series was more than twice the following term. See Zimmermann [1, pp. 47–48]. Once i is so chosen the conclusion follows from the term-by-term comparison with the geometric series $\Sigma\ 1/2^k$. On the use of the comparison with this series to insure the good behavior of infinite series see Newton [4] (also in Newton [2, vol. 1, p. 24]) and Euclid, *Elements*, book X, proposition 1. For evidence that Lagrange knew Arbogast's unpublished memoir, see his own statement in *Fonctions analytiques* (Lagrange [16, p. 5]; also in Lagrange [9, vol. 9, p. 19]).

16. *Fonctions analytiques* (Lagrange [9, vol. 9, p. 29]).

17. Lagrange [9, vol. 9, p. 29]. Compare Euler's analogous remark in Euler [3, section 122], quoted earlier, on p. 117. Compare also Lagrange, *Calcul des fonctions* (Lagrange [4]; in Lagrange [9, vol. 10, p. 101]. For Lagrange's own "applications" see later in the chapter.

18. See chapter 2 for a full discussion.

19. *Calcul des fonctions* (Lagrange [9, vol. 10, pp. 86–87]). Compare *Fonctions analytiques* (Lagrange [9, vol. 9, pp. 72, 77]). Lagrange there gave this alternative form: $f(x+i) = f(x) + if'(x) + i^2Q$, where Q is finite.

20. In *Fonctions analytiques* (1797) Lagrange derived the Lagrange property from Euler's criterion. In the *Calcul des fonctions* (1801) he derived the Lagrange property of the derivative directly from the existence of the Taylor series of a function and then proved Euler's criterion as a corollary of Taylor's theorem with Lagrange remainder (Lagrange [9, vol. 10, pp. 100–101]). The earlier derivation strongly supports my conclusion that Euler's criterion led Lagrange to the crucial property of the derivative,

while the later one shows that he recognized the equivalence of the two properties, provided the function $f(x)$ is represented by its Taylor series.

21. *Calcul des fonctions* (Lagrange [9, vol. 10, p. 87]).

22. *Calcul infinitésimal* (Cauchy [14, series 2, vol. 4, pp. 44–46]).

23. For the central role of (5.6) in Cauchy's calculus see *Calcul infinitésimal* (Cauchy [14, series 2, vol. 4, pp. 89, 123, 131, 151–152, 243]) and see later in this chapter.

24. The relationship between the problems treated in these papers has often been noted. See, for example, Pringsheim and Molk [1]. For the history of the proof techniques and the logical relationship between them see Grabiner [4]. Excerpts from the relevant text and a connecting narrative may be found in Dugac [2].

25. See chapter 3.

26. *Calcul des fonctions*: 1st ed., 1801; 2nd ed., 1806 (Lagrange [9, vol. 10, p. 86]). Compare *Fonctions analytiques* (Lagrange [9, vol. 9, pp. 78–80]): "If a prime function of x, like $f'(x)$, is always positive for all values of x from $x = a$ to $x = b$, b greater than a, then the difference of primitive functions corresponding to these two values of x, that is, $f(b) - f(a)$, necessarily will be positive." The version in *Calcul des fonctions* has been chosen because the proof there is better, specifically in deriving the relevant inequalities.

27. *Calcul des fonctions* (Lagrange [9, vol. 10, p. 87]). In the version of the lemma given in the *Fonctions analytiques* Lagrange did not make entirely clearly whether or not the quantity the *Calcul des fonctions* calls V had to be restricted to positive values; see Lagrange [9, vol. 9, pp. 78–80]. No confusion on this point remains in the *Calcul des fonctions*.

28. As we have observed in chapter 3, even as late as d'Alembert's treatment of the binomial series (1768) people easily could be led into errors by not distinguishing whether the terms of a series or their absolute values are small. For another sample of the eighteenth-century confusion on this point see "Can a Variable Surpass Its Limit?" in chapter 4, p. 84ff.

29. By *limit* Lagrange here meant *bound*; *Calcul des fonctions* (Lagrange [9, vol. 10, p. 88]). Recall that the derivatives $f'(x + ki)$ are all assumed finite.

30. The confusion between *greater than zero* and *bounded away from zero* is frequent in the eighteenth century; the distinction was first correctly made (in practice, though not in words) by Cauchy.

31. Because, though he does not say so, for such D,

$$0 < [if'(x) + \cdots + f'(x + \{n-1\}i)] - niD$$

and

$$[if'(x) + \cdots + if'(x + \{n-1\}i)] + niD$$
$$< 2[if'(x) + \cdots + if'(x + \{n-1\}i)].$$

32. For Lagrange, *primitive functions* were the functions whose derivatives were being considered. Finding a primitive function, for Lagrange, was what most eighteenth-century mathematicians would call *integration.*

33. *Calcul des fonctions* (Lagrange [9, vol. 10, p. 91]). Compare *Fonctions analytiques* (Lagrange [9, vol. 9, pp. 80–81]). Lagrange does not write $\leqslant$ but $<$; however, the context indicates that he means $\leqslant$.

34. *Calcul des fonctions* (Lagrange [9, vol. 10, pp. 91–95]); compare *Fonctions analytiques* [9, vol. 10, pp. 80–85]. Again, Lagrange did not use the notation $\leqslant$, but $<$.

35. For Lagrange's attempt to prove this, see chapter 3, p. 73.

36. Ampère [1].

37. For explicit references to *Fonctions analytiques* see Ampère [1, pp. 160, 169]; to *Calcul des fonctions* see Ampère [1, p. 163]; for references to the common doctrine of both Lagrange's books see Ampère [1, pp. 149, 165, *et passim*].

38. Cauchy [14, series 2, vol. 3, pp. vii–viii].

39. *Cours d'analyse* (Cauchy [14, series 2, vol. 3, p. viii]) and *Applications du calcul infinitésimal à la géometrie* (Cauchy [14, series 2, vol. 5, p. 10]).

40. *Calcul infinitésimal* (Cauchy [14, series 2, vol. 4, p. 44n]). He mentioned the paper again in *Calcul différentiel* (Cauchy [14, series 2, vol. 4, p. 268]).

41. See *Calcul différentiel* (Cauchy [14, series 2, vol. 5, p. 268]) and *Calcul infinitésimal* (Cauchy [14, series 2, vol. 4, pp. 9–10]), for instance.

42. *Fonctions analytiques* (Lagrange [9, vol. 9, p. 85]). The reference, without any evaluative comment by Lagrange, links Ampère [1] to leçon IX of the *Calcul des fonctions* (Lagrange [9, vol. 10, pp. 85–105]). I wish to thank Lagrange for this reference to Ampère.

43. The major source of this interpretation seems to be Pringsheim and Molk [1, p. 44].

44. Ampère [1, p. 149].

45. Ampère introduced this (Ampère [1, p. 156]) by saying it was "a definition of the derived function $f'(x)$ that seems to me the most general and the most rigorous one possible." Cauchy's theorem (5.5) proves that $f'(x)$ satisfies Ampère's definition.

46. In fact, if the derivative $f'(x)$ is continuous, then no other continuous function has this property. Ampère's uniqueness proof, with a bit of modification, can be adapted to show this. Ampère himself, however, did not explicitly restrict himself to continuous functions; before Cauchy, the distinction between continuous and noncontinuous functions would not have seemed important in this context.

47. Ampère [1, p. 149].

48. Ampère [1, pp. 154–155]. This is Ampère's notation. He did not mention Lagrange explicitly as the source of this particular property. See, however, note 37 in this chapter.

49. Ampère [1, p. 151].

50. *Cours d'analyse* (Cauchy [14, series 2, vol. 3, note II, theorem XII, p. 368]). See the appendix for a translation of Cauchy's result. Cauchy did not acknowledge Ampère when giving his own result. He probably was not conscious of the relationship; see Freudenthal [1] for a discussion of Cauchy's way of working.

51. Ampère [1, pp. 151–154].

52. Ampère [1, pp. 154–155]. Compare Lagrange's characterization of the function iP as going to zero with i in *Fonctions analytiques* (Lagrange [9, vol. 9, pp. 28–29]) and also p. 95.

53. Ampère [1, p. 152] expressed these inequalities verbally, saying "plus grande," "plus petite," but said specifically that he meant "greater than or equal to," or "less than or equal to." Additionally, I have substituted the notation $[a, k]$ for his verbal description.

54. I have tried to isolate the theorem of interest. But part of the difficulty in reading Ampère's paper is that the proof just considered is embedded in the proof of the theorem that $f'(x)$ cannot be zero or infinite on the whole interval. Of course (5.8) does imply this.

55. Ampère [1, p. 162].

56. *Calcul infinitésimal* (Cauchy [14, series 2, vol. 4, p. 44]. Cauchy said "comprise" for what I have translated as "included"; he meant $\leqslant$. For strict inequality, he said "renfermé," which I have translated as "lying between." Lagrange did not make this dis-

tinction clear in his proofs; Ampère at least tried to. I use $|i|$ for Cauchy's "numerical value" of i.

57. Cauchy did not refer to Ampère, but to his own result in the *Cours d'analyse*, note II.

58. See the works on the calculus cited in chapter 2. A selection may be consulted in Struik [1, for instance pp. 272–284 (Leibniz); 303–309 (Newton); 312, 315 (L'Hôpital); 400–401 (Euler)]. But recall the contrary example of Euler's treatment of extrema.

59. *Leçons sur les applications du calcul infinitésimal à la géométrie* (1826) (Cauchy [5]; also in Cauchy [14, series 2, vol. 5, p. 9]). This work was intended by Cauchy as a companion volume to the *Calcul infinitésimal* of 1823.

60. Cauchy [5] (also in Cauchy [14, series 2, vol. 5, p. 44]). I have deleted italics from the words "tangent" and "touches."

61. Cauchy [5] (also in Cauchy [14, series 2, vol. 5, p. 44]).

62. For instance, see *Calcul infinitésimal* (Cauchy [14, series 2, vol. 4, pp. 88–89, 217]). The notation is Cauchy's, save for $0 \leqslant \theta \leqslant 1$, which he expressed verbally.

63. Bolzano also adapted some of Lagrange's arguments of this type. See, for instance *Functionenlehre* (Bolzano [4, vol. 1, p. 155ff]). For an explicit acknowledgement by Bolzano of Lagrange's prior use of such techniques in Lagrange's *Fonctions analytiques* and *Calcul des fonctions*, see Bolzano, *Functionenlehre* (Bolzano [2, vol. 1, p. 170]).

64. Cauchy, again like Lagrange, chose first to expound the calculus analytically, and then to give a separate, almost equally lengthy, treatment of its applications to geometry. Lagrange carried out these separate tasks in the separate parts I and II of his *Fonctions analytiques*; Cauchy did it in two separate books, his *Calcul infinitésimal* of 1823 and his *Leçons sur les applications du calcul infinitésimal à la géométrie* of 1826. In this separation of treatments, Lagrange in turn was following the practice of Euler in volumes 1 (analytic) and 2 (geometric) of his *Introductio in analysin infinitorum* of 1748. In part III of *Fonctions analytiques* Lagrange discussed the application of the calculus to mechanics. But in his textbooks from the Ecole Polytechnique Cauchy did not devote space to this specific set of problems, though of course he worked with convergent series in mechanics and was proficient in mathematical physics.

65. For Euler, see pp. 117–118. For Lagrange, see *Fonctions analytiques* (Lagrange [9, vol. 9, pp. 233–237]). For a modern

treatment of maxima and minima using Lagrange's procedures see, for example, D. Widder, *Advanced Calculus* (New York: Prentice-Hall, 1947), pp. 76–77, 99.

66. *Calcul infinitésimal* (Cauchy [14, series 2, vol. 4, pp. 88–90]).

67. *Calcul infinitésimal* (Cauchy [14, series 2, vol. 4, pp. 90–92]).

68. *Fonctions analytiques* (Lagrange [9, vol. 9, pp. 183–189]).

69. Compare Euclid *Elements*, book III, definition 1 and proposition 16, for the case of the circle. This definition of tangent in general is found before Lagrange: see Maclaurin [2, book I, p. 179].

70. This theory began by proving the following: Given two curves represented by $f(x)$ and $F(x)$ intersecting at x such that $f'(x) = F'(x)$, another curve $\phi(x)$ can be interposed between the two given curves only if $f'(x) = F'(x) = \phi'(x)$ at the fixed point x. Lagrange proved this by applying Taylor's theorem with Lagrange remainder to the differences $f(x + i) - F(x + i)$ and $f(x + i) - \phi(x + i)$, and by careful computation with the inequalities resulting from interpreting the statement "ϕ lies between the curves f and F" and those Taylor-series remainders. Analogous results were derived when $f''(x) = F''(x)$, and so forth. These results led Lagrange to define two curves as having a contact of order m at a point when they had m equal derivatives at that point; see *Fonctions analytiques* (Lagrange [9, vol. 9, pp. 189, 198]). The detailed study and theory of orders of contact of curves is considered to begin with Lagrange; see Struik, *Lectures on Classical Differential Geometry* (Cambridge, Mass.: Addison-Wesley, 1950), p. 25.

71. Cauchy [5] (also in Cauchy [14, series 2, vol. 5, pp. 77–80]).

72. *Fonctions analytiques* (Lagrange [9, vol. 9, pp. 193–196]), Cauchy [5, leçon 7] (also in Cauchy [14, series 2, vol. 5, especially pp. 115–116]).

73. *Fonctions analytiques* (Lagrange [9, vol. 9, pp. 229–231]).

74. *Fonctions analytiques* (Lagrange [9, vol. 9, pp. 151–157]).

75. *Fonctions analytiques* (Lagrange [9, vol. 9, pp. 258–267]). For Cauchy see *Calcul infinitésimal* (Cauchy [14, series 2, vol. 4, pp. 47–49, 221–222]).

76. Suppose two plane curves are tangent to each other at the point P. Let i be infinitely small. Choose a point Q on the first curve and R on the second such that $PQ = PR = i$. Let w represent the angle between the lines PQ and PR. Since the curves are tangent, w, like i, is infinitely small. Cauchy then defined the

order of contact a between the curves as the order of the infinitesimal w considered as a function of i; see Cauchy [5] (also in Cauchy [14, series 2, vol. 5, p. 138]). Cauchy was proud of this definition because, among other things, it defined orders of contact between curves independently of the coordinate system; see "Avertissement" in Cauchy [5] (also in Cauchy [14, series 2, vol. 5, pp. 9–10]).

77. For Cauchy's theory of infinitesimals of various orders, which is based on his theory of limits, see *Cours d'analyse* (Cauchy [14, series 2, vol. 3, p. 38ff]) and compare Cauchy [5] (also in Cauchy [14, series 2, vol. 5, pp. 132–137]).

78. Cauchy [5] (also in Cauchy [14, series 2, vol. 5, p. 138]).

79. *Calcul infinitésimal* (Cauchy [14, series 2, vol. 4, pp. 243–244]).

80. *Calcul infinitésimal* (Cauchy [14, series 2, vol. 4, pp. 246–247]) for L'Hôpital's rule; he also applied (5.14) to derive some results on orders of infinitesimals and Taylor's theorem; see *Calcul infinitésimal* (Cauchy [14, series 2, vol. 4, pp. 257–261]). For Cavalieri's theorem, see Cauchy [5] (also in Cauchy [14, series 2, vol. 5, p. 457]).

81. To within an alphabetical isomorphism. Lagrange used D (for *donnée*) for our epsilon, i (for *indeterminée*) both for his increment and for what we call delta.

82. Except, of course, for his not distinguishing between convergence and uniform convergence, a distinction not made clearly until the 1840s.

Chapter 6

1. For instance, see Euler [4, p. 1] (also in Euler [6, series 1, vol. II, p. 11]) and Lacroix [1, vol. 2, pp. 1–2]. This point has been made and amply documented by numerous historians, for example, Vivanti [1, pp. 741–742], and Voss [1, especially pp. 88–89].

2. Cajori [2, vol. 2, p. 243] reproduces the page of Leibniz's manuscript of 29 October 1675 in which the integral sign first appeared. The first publication of the integral sign was in Leibniz [1]; an excerpt from this paper may be found in Struik [1, pp. 281–282]. On Leibniz's integral see also Baron [1, p. 284].

3. See Euler [4, p. 4].

4. Lebesgue [1, p. 5]. I owe this reference to Iushkevich [5].

5. Postscript, 1823, added in 1822 to a paper begun in 1821, "Mémoire sur l'intégration des équations lineares aux differentielles partielles et à coefficients constants" (Cauchy [11]). The paper is in *Oeuvres complètes d'Augustin Cauchy* (Cauchy [14, series 2, vol. 1, pp. 275–357]): pp. 275–333 comprise the body of the

paper; pp. 333–354 is an addendum called "Observations générales et additions," which—because of a reference to the *Calcul infinitésimal* on p. 334—we know was composed *after* he had conceived the theory of the definite integral as a sum; the postscript is found on pp. 354–357, and apparently was composed after the "Observations générales." The point I have quoted is on p. 354 of Cauchy [14, series 2, vol. 1].

6. Iushkevich [5, p. 406].

7. This point has been argued at length by Jourdain [2, especially p. 665; compare pp. 679, 681–682, 684–685]. Compare Grattan-Guinness [3, pp. 44–45].

8. Fourier [2, for example sections 186, 220]. Compare references in note 7.

9. Legendre [1, pp. 309–314]. Compare Grattan-Guinness [3, p. 38], who gives another example from elsewhere in Legendre's work.

10. Cauchy [9] (also in Cauchy [14, series 1, vol. 1, pp. 319–506]).

11. *Calcul infinitésimal* (Cauchy [14, series 2, vol. 4, pp. 140–144]). Compare Cauchy [9], [14, series 1, vol. 1, pp. 402–406].

12. Cauchy himself suggested that complex integration helped motivate his new definition in the postscript cited in note 5. In 1825 Cauchy applied his sum definition to complex integrals in Cauchy [10], not published, however, until 1874–1875. An English translation of this 1825 memoir may be found in Iacobacci [1]. I shall return to Cauchy's 1825 paper at the end of this chapter.

13. Cauchy [9] (also in Cauchy [14, series 1, vol. 1, p. 337ff]). For an account of this paper see Iushkevich [5, p. 389ff] and compare Iacobacci [1, p. 350ff].

14. Cauchy [9] (also in Cauchy [14, series 1, vol. 1, pp. 329–330]).

15. Cauchy [14, series 1, vol. 1, p. 322].

16. Cauchy [14, series 1, vol. 1, pp. 338, 390].

17. Letter to Bessel of 1811; often cited, for example by Kline [1, p. 632].

18. Poisson [1]. For references by Cauchy to Poisson's work see, for example, Cauchy [14, series 1, vol. 1, pp. 333–334, 354ff]. For their relation see Iushkevich [5, pp. 397–401] and later in this chapter.

19. For instance, Poisson noted that $\int_{-1}^{1} dx/x$ should by symmetry considerations be zero, yet if $\int_{-1}^{1} dx/x$ is evaluated over a

complex path, then transforming it by the substitution $x = \cos z + i \sin z$ yields $-(2n+1)\pi$ for the definite integral. See Poisson [1, pp. 320–321].

20. Cauchy [10], cited in note 12.

21. Cauchy [14, series 1, vol. 1, pp. 333–334]; italics mine. As pointed out in note 5, this was written closely following *Calcul infinitésimal* and published in the same year—1823.

22. Postscript, in Cauchy [14, series 1, vol. 1, p. 354].

23. An account of the work of Euler, Lacroix, and Poisson, and of some of the resemblances between their work and Cauchy's, is given in Iushkevich [5]. Iushkevich also treats there Cauchy's work on complex integration and on improper integrals. This study deserves to be more widely known. I shall cite Iushkevich in subsequent footnotes whenever his findings have anticipated my own. I go beyond Iushkevich in stressing how Lagrange showed Poisson and Cauchy how to use the Taylor-series remainders, in drawing the parallel between Cauchy's existence arguments here and in his other adaptations of eighteenth-century approximations, in my analysis of Cauchy's 1825 memoir (for which see later in the chapter), and in my analysis of some of Cauchy's proofs.

24. *Calcul infinitésimal* (Cauchy [14, series 2, vol. 4, p. 125]).

25. *Calcul infinitésimal* (Cauchy [14, series 2, vol. 4, p. 123]).

26. *Calcul infinitésimal* (Cauchy [14, series 2, vol. 4, p. 125]).

27. See chapter 4, and *Cours d'analyse* (Cauchy [14, series 2, vol. 3, pp. 115–116]).

28. *Calcul infinitésimal* (Cauchy [14, series 2, vol. 4, p. 125]); italics his.

29. *Calcul infinitésimal* (Cauchy [14, series 2, vol. 4, p. 126]).

30. Euler [4] (also in Euler [6, series 1, vol. 11, p. 184]). Rewriting (6.5) in Cauchy's notation, let the "constant of integration" b be zero, let the function X be represented by $f(x)$, let $a = x_0$, $a' = x_1, \ldots, 'x = x_{n-1}$, and $x = X$. Then (6.5) becomes $y = f(x_0)(x_1 - x_0) + f(x_1)(x_2 - x_1) + \cdots + f(x_{n-1})(X - x_{n-1})$, which is exactly (6.1). Compare Iushkevich [5, p. 378].

31. Euler [4] (also in Euler [6, series 1, vol. 11, p. 186]). Compare Iushkevich [5, p. 379]. Euler showed that the integral is always included between

$$b + A(a' - a) + A'(a'' - a) + \cdots + 'X(x - 'x)$$

and

$$b + A'(a' - a) + A''(a'' - a) + \cdots + X(x - 'x).$$

Euler does not use the word *monotonic.*

32. Euler [6, series 1, vol. 11, pp. 186–187].

33. Legendre [1, part III, p. 314]. For the derivation see his p. 309ff. For Cauchy's knowledge of Legendre's work, though without reference to this particular passage see, for example, Grattan-Guinness [3, pp. 36–41]. For the Cauchy principal value as a way of dealing with singular points, see Cauchy [9] (also in Cauchy [14, series 1, vol. 1, pp. 319–506]).

34. 1822. See, for example, articles 186, 346, 415–418; compare Grattan-Guinness [3, p. 44].

35. Lacroix [1, vol. 2, p. 137]. This is exactly Euler's expression (6.5) save for a change in notation. Lacroix expressly referred to Euler's *Institutiones calculi integralis* as his source in writing this section; Lacroix [1, vol. 2, p. 135]. However, he derived (6.6) somewhat differently; see Iushkevich [5, p. 381].

36. Lacroix [1, vol. 2, p. 139ff].

37. Lacroix wrote $Y' = X(a)$, $Y_1{}' = X(a_1)$.

38. Lacroix [1, vol. 2, p. 140].

39. Lacroix [1, vol. 2, p. 140].

40. *Calcul infinitésimal* (Cauchy [14, series 2, vol. 4, p. 129]).

41. Cauchy [14, series 2, vol. 4, pp. 131–133], for instance; compare the pages in Lacroix [1] that I have just discussed.

42. Lacroix [1, vol. 2, p. 140]: "Other cases are reducible to the two we have just considered." Lacroix did not, however, make any explicit appeal to the concept of continuous function.

43. Poisson [1, pp. 295–341]. For Cauchy's acquaintance with this paper, see his "P.S." to Cauchy [11, p. 354].

44. He credits the fact that the integral is a sum to Lacroix [1, vol. 2, article 471]; Poisson [1, p. 319].

45. Poisson [1, p. 319].

46. Poisson [1, p. 323]. Iushkevich [5, p. 399] cites this also.

47. Euler [4] (also in Euler [6, series 1, vol. 11, p. 199]). Compare Lacroix, [1, vol. 2, pp. 136–139], where such approximations are discussed in both the equal and unequal-division cases.

48. Poisson [1, p. 322].

49. Poisson [1, p. 323]. I use absolute value signs; he said "abstraction made of the sign." Also, he wrote $<$ where I have $\leqslant$; the

latter seems to be intended. He expressed $R_0 + R_1 + \cdots + R_{n+1} = T$, and omitted the step $|F(b) - f(a) - S| \leqslant |nM(\alpha^{1+k})|$.

50. Poisson [1, p. 323]. The proof is reproduced by Iushkevich [5, p. 399].

51. See Lagrange [16] (in Lagrange [9, vol. 9, pp. 80–81]), Lacroix [1, vol. 2, p. 141]. Poisson's knowledge of these works has been touched on in this chapter.

52. Poisson [1, pp. 329–330]; italics mine.

53. Cauchy [15, pp. 151–152]. Compare Moigno's version, in Birkhoff [1, pp. 8–11].

54. Cauchy [15, p. 131, formula 19].

55. For Cauchy's source for the mean-value theorem for integrals, see Lagrange, *Fonctions analytiques* (Lagrange [9, vol. 9, p. 81]) and Lacroix [1, vol. 2, pp. 141–142], where Lacroix follows Lagrange's notation.

56. *Fonctions analytiques* (Lagrange [9, vol. 9, pp. 238–239]). He also treated the case where f decreases. I use the notation $[x, x + i]$ for the interval; he used words only. For (6.16), Lagrange, characteristically, did not draw a diagram; the reader may easily supply one. I have substituted , which is meant, for his usual (and ambiguously used) $<$.

57. *Fonctions analytiques* (Lagrange [9, vol. 9, p. 239]). I have supplied the absolute value signs; he used the concept, but not the notation.

58. See note 55.

59. "Avertissement," Cauchy [4] (in Cauchy [14, series 2, vol. 4, p. 268]), cites the *Institutiones calculi differentialis* (Euler [3]); the *Institutiones calculi integralis* (Euler [4]) is cited in Cauchy [8], also in Cauchy [14, series 2, vol. 1, p. 512].

60. Cauchy made no explicit reference here to Euler; as usual, I argue from the close resemblance. For Euler's treatment of this approximation, see Euler [4, p. 493] or Euler [6, vol. 11, pp. 424–425]. The standard historical account of this method, pointing out the almost certain debt of Cauchy to Euler, is in Painlevé [1, especially p. 193, note 5]. Cauchy's method was first published as Cauchy [19]; I have not seen this edition. It was reprinted in 1840 as *Exercices d'analyse*, vol. 1, and may be found in Cauchy [14, series 2, vol. 11, p. 399ff].

61. Cauchy [5, pp. 431–432].

62. Cauchy [5, pp. 443–446].

63. Cauchy [5, pp. 464–466].

64. Cauchy [5, pp. 498–499].

65. Cauchy [10, section 1], given in Iacobacci [1, p. 460].

66. Cauchy [10, section 2], cited in Iacobacci [1, p. 462]. Compare Birkhoff [1, p. 33].

67. Cauchy [10, section 2], cited in Iacobacci [1, p. 463]; Birkhoff [1, p. 34].

68. Cauchy [10, section 3], cited in Iacobacci [1, p. 464]; Birkhoff [1, p. 34, 34n].

69. See Cauchy [9] (in Cauchy [14, series 1, vol. 1, p. 329ff, especially pp. 402–404]).

70. See Riemann [2] (in Riemann [1, especially pp. 239–241]). For a Riemann reference to Cauchy, see Riemann [1, p. 234], and for the chain of influence from Cauchy to Dirichlet to Riemann, see Hawkins [1, chapter 1] and compare Dauben [1, pp. 11–14].

References

N. H. Abel

[1] *Oeuvres complètes*, 2 vols., eds. L. Sylow and S. Lie, Christiana: Grondahl & Son, 1881.

[2] Recherches sur la série $1 + (m/1)x + (m(m-1)/1 \cdot 2)x^2 + (m(m-1)(m-2)/1 \cdot 2 \cdot 3 \cdot x^3 + \cdots$, *Crelles Journal* 1(1826). In Abel [1, pp. 221–250]. German version, Leipzig: Engelmann, Ostwalds Klassiker 71, 1895.

J. d'Alembert

[1] Différentiel. In J. d'Alembert, C. Bossut, J. J. deLalande, J. M. Marquis de Condorcet [1].

[2] Fluxion. In J. D'Alembert, C. Bossut, J. J. deLalande, J. M Marquis de Condorcet [1].

[3] Réflexions sur les suites et sur les racines imaginaires. In *Opuscules mathématiques*, vol. 5, Paris: Briasson, 1768, pp. 171–215.

[4] Sur les principes métaphysiques du calcul infinitésimal. In *Mélanges de littérature, d'histoire, et de philosophie*, vol. 5, Amsterdam: Chatelan, 1767.

J. d'Alembert, C. Bossut, J. J. deLalande, J. M. Marquis de Condorcet

[1] *Dictionnaire encylopédique des mathématiques*, Paris: Hotel de Thou, 1789 (the mathematical articles from the Diderot–d'Alembert *Encyclopédie*).

J. d'Alembert and de la Chapelle

[1] Limite. In J. D'Alembert, C. Bossut, J. J. deLalande, J. M. Marquis de Condorcet [1].

A. M. Ampère

[1] Recherches sur quelques points de la théorie des fonctions derivées qui conduisent à une nouvelle démonstration de la série de Taylor, et à l'expression finie des terms qu'on néglige lorsqu'on arrête cette série à un terme quelconque, *Journal de l'Ecole Polytechnique*, Cahier 13, 6(1806) : 148–181.

L. F. A. Arbogast

[1] *Du calcul des dérivations*, Strasbourg: Levrault Frêres, An VIII (1800).

[2] Essai sur de nouveaux principes de calcul différentiel et intégral, independans de la théorie des infiniment-petits et celle des

limites, Biblioteca Medicea-Laurenziana, Florence, MS Codex Ashburnham Appendix sig. 1840.

[3] *Mémoire sur la nature des fonctions arbitraires qui entrent dans les integrales des équations aux differences partielles*, St. Petersburg: Académie Impériale des Sciences, 1791.

Archimedes
[1] *The Works of Archimedes*, ed. T. L. Heath. Cambridge: 1912. Reprinted New York: Dover, n.d.

M. Baron
[1] *Origins of the Infinitesimal Calculus*, Oxford: Pergamon Press, 1969.

E. T. Bell
[1] *The Development of Mathematics*, New York: McGraw-Hill, 1945.

J. Ben-David
[1] *The Scientist's Role in Society*, Englewood Cliffs: Prentice-Hall, 1971.

G. Berkeley
[1] *A Defence of Freethinking in Mathematics*, 1735. In *The Works of George Berkeley*, eds. A. A. Luce and T. R. Jessop, vol. 4, London: T. Nelson, 1951.

[2] *The Analyst, or a Discourse Addressed to an Infidel Mathematician*, 1734. In *The Works of George Berkeley*, eds. A. A. Luce and T. R. Jessop, vol. 4, London: T. Nelson, 1951.

D. Bernoulli
[1] Observationes de seriebus quae formantur ex additione vel subtractione quacunque terminorum se mutuo consequentium ..., *Commentarii academiae scientiarum imperialis ... petropolitanae* 3(1728):85–100.

Jakob Bernoulli
[1] *Opera*, 2 vols., Geneva: Cramer, 1744.

Johann Bernoulli
[1] *Die Differentialrechnung aus dem Jahre 1691/92*, Ostwalds Klassiker 211, Leipzig: Engelmann, 1924.

[2] *Opera omnia*, 4 vols., Lausanne and Geneva: Bousquet, 1742. Reprinted Hildesheim: Olms, 1968–.

K. R. Biermann
[1] Weierstrass, *Dictionary of Scientific Biography*, vol. 14, New York: Scribner's, 1976, pp. 219–224.

G. Birkhoff

[1] ed., *A Source Book in Classical Analysis*, Cambridge, Massachusetts: Harvard University Press, 1973.

B. Bolzano

[1] *Die binomische Lehrsatz*, Prague: Enders, 1816.

[2] *Functionenlehre*. In Bolzano [4, vol. 1].

[3] *Rein analytischer Beweis des Lehrsatzes dass zwischen je zwey [sic] Werthen, die ein entgegengesetztes Resultat gewaehren, wenigstens eine reele Wurzel der Gleichung liege*, Prague: 1817. Reprinted in Ostwalds Klassiker 153, ed. P. E. B. Jourdain, Leipzig: Engelmann, 1905. French translation: Démonstration purement analytique du theorème: entre deux valeurs quelconques qui donnent deux résultats de signes opposés se trouve au moins une racine réele de l'équation, *Révue d'histoire des sciences* 17(1964):136–164. English translation Russ [1].

[4] *Schriften*, Prague: Königlichen Böhmischen Gesellschaft der Wissenschaften, 1930.

N. Bourbaki

[1] *Elements d'histoire des mathématiques*, Paris: Hermann, 1960.

C. Boyer

[1] *History of the Calculus and Its Conceptual Development*, New York: Dover [reprint], 1959.

[2] *History of Mathematics*, New York: Wiley, 1967.

A. Brill and M. Noether

[1] Die Entwicklung der Theorie der algebraischen Funktionen in älterer und neuer Zeit, *Jahresbericht der Deutschen Mathematische Vereinigung*, 3(1894):107–566.

F. Cajori

[1] *A History of the Conceptions of Limits and Fluxions in Great Britain from Newton to Woodhouse*, Chicago and London: Open Court, 1919.

[2] *History of Mathematical Notations*, 2 vols., Chicago: Open Court, 1928–1929.

G. Cantor

[1] Über die Ausdehnung einer Satzes aus der Theorie der trigonometrisches Reihen, *Mathematische Annalen* 5(1872):123–132, in G. Cantor [2, 92–102].

[2] *Gesammelte Abhandlungen*, Berlin: 1932, Hildesheim: G. Olms, 1962.

M. Cantor

[1] *Vorlesungen über Geschichte der Mathematik*, 4 vols., Leipzig: Teubner, 1894–1901.

L. N. M. Carnot

[1] *Réflexions sur la métaphysique du calcul infinitésimal*, Paris: Duprat, 1797.

[2] *Réflexions sur la métaphysique du calcul infinitésimal*, revised and expanded, Paris: Courcier, 1813.

A. L. Cauchy

[1] *Cours d'analyse de l'école royale polytechnique. 1re partie: analyse algébrique* [all published], Paris: 1821. In Cauchy [14, series 2, vol. 3]; all page references are to this edition.

[2] *Dei metodi analitici*, Roma: 1843 (Italian memoir, *not* a translation of Cauchy [1]).

[3] *Exercices d'analyse et de physique mathématique*, 4 vols., Paris: 1840–1842. In Cauchy [14, series 2, vols. 11–14].

[4] *Leçons sur le calcul différentiel*, Paris: de Bure Frères, 1829. In Cauchy [14, series 2, vol. 4, pp. 263–609]; all page references are to this edition.

[5] *Leçons sur les applications du calcul infinitésimal à la géometrie*, Paris: 1826–1828. In Cauchy [14, series 2, vol. 5]; all page references are to this edition.

[6] *Lehrbuch der algebraischen Analysis*, tr. C. L. B. Huzler, Königsberg: Bornträger, 1828 (German translation of *Cours d'analyse*).

[7] Mémoire d'analyse infinitésimale, *Comptes rendus* (1843). In Cauchy [14, series 1, vol. 8, pp. 11–17].

[8] Mémoire sur divers formules rélatives à la théorie des intégrales définies, *Journal de l'Ecole Polytechnique*, Cahier 28, 17(1815): 147ff. In Cauchy [14, series 2, vol. 1, pp. 467–567].

[9] Mémoire sur les intégrales définies, *Mémoires des divers savants*, series 2, 1(1827): 601–799 (written 1814). In Cauchy [14, series 1, vol. 1, pp. 329–506].

[10] Mémoire sur les intégrales définies prises entre des limites imaginaires, *Bulletin des sciences mathématiques*, series 1, 7(1874): 265–304; 8(1875): 43–55, 148–159 (written 1825). German translation ed. P. Staeckel, Leipzig: Engelmann, Ostwalds Klassiker 112, 1900.

[11] Mémoire sur l'intégration des équations lineares aux différentielles partielles et à coefficients constantes, *Journal de l'Ecole Polytechnique*, Cahier 19, 12(1823): 511ff. In Cauchy [14, series 2, vol. 1, pp. 275–357].

[12] Note sur les séries convergentes dont les divers termes sont des fonctions continués d'une variable reélle ou imaginaire entre des

limites données, *Comptes rendus*, 36(1853). In Cauchy [14, series 1, vol. 12, pp. 30–36].

[13] *Nouveaux exercices des mathématiques*, Prague: 1835. In Cauchy [14, series 2, vol. 10, pp. 185–464].

[14] *Oeuvres complètes d'Augustin Cauchy*, publieés sous la direction scientifique de l'Académie des Sciences, Paris: Gauthier-Villars, series 1, 12 vols., series 2, 15 vols. so far; 1882–.

[15] *Résumé des leçons données à l'école royale polytechnique sur le calcul infinitésimal*, vol. 1 [all published], Paris: Imprimérie Royale, 1823. In Cauchy [14, series 2, vol. 4, pp. 5–261]; all page references are to this edition.

[16] Sur la développement des fonctions en séries, et sur l'intégration des équations differentelles ou aux differences partielles, *Bulletin de la Societé Philomathique* (1822): 49–54. In Cauchy [14, series 2, vol. 2, pp. 276–282].

[17] Sur la plus grande erreur à craindre dans un result moyen, et sur le système de facteurs qui rend cette plus grande erreur un minimum, *Comptes rendues* 37(1853). In Cauchy [14, series 1, vol. 12, pp. 114–124].

[18] Sur les intégrales définies prises entre des limites imaginaires, *Bulletin de Ferussac* 3(1825): 214–221. In Cauchy [14, series 2, vol. 2, pp. 57–65].

[19] *Sur l'intégration des équations différentielles*, Prague: 1835. Reprinted in *Exercices d'analyse*, 1840. In Cauchy [14, series 2, vol. 11, pp. 399ff].

[20] *Vorlesungen über die Anwendungen der Infinitesimalrechnung auf die Geometrie* tr. C. Schnuse, Braunschweig: G. C. E. Meyer, 1840 (German translation of *Leçons sur les applications du calcul infinitésimal à la géometrie*).

[21] *Vorlesungen über die Differenzialrechnung*, tr. C. H. Schnuse, Braunschweig: G. Meyer, 1836 (German translation of the *Leçons sur le calcul différentiel*).

J. M. Marquis de Condorcet

[1] Approximation. In J. D'Alembert, C. Bossut, J. J. deLalande, J. M. Marquis de Condorcet [1].

[2] Série ou suite. In J. d'Alembert, C. Bossut, J. J. deLalande, J. M. Marquis de Condorcet [1].

J. Dauben

[1] *Georg Cantor*, Cambridge, Massachusetts: Harvard University Press, 1978.

R. Dedekind
[1] *Stetigkeit und die Irrationalzahlen*, 1872. Reprinted as *Continuity and Irrational Numbers*, in R. Dedekind, *Essays on the Theory of Numbers*, New York: Dover, 1963.

J. B. Delambre
[1] Notice sur la vie de Lagrange. In *Oeuvres de Lagrange* ([Lagrange 9, vol. 1, pp. viii–li]).

S. Dickstein
[1] Zur Geschichte der Prinzipien der Infinitesimalrechnung. Die Kritiker der "Théorie des fonctions analytiques" von Lagrange, *Abhandlung zur Geschichte der Mathematik* 9(1899): 65–79.

D. Diderot
[1] De la interprétation de la nature. In D. Diderot, *Oeuvres philosophiques*, ed. P. Vernière, Paris: Garnier, 1961.

J. Dieudonné
[1] *Abrégé d'histoire des mathématiques, 1700–1900*, 2 vols., Paris: Hermann, 1978.

P. G. L. Dirichlet
[1] *Werke*, 2 vols., ed. L. Kronecker, Berlin: 1889–1897. Reprinted New York: Chelsea, 1969.

J. Dubbey
[1] Cauchy's contribution to the establishment of the calculus, *Annals of Science* 22(1966): 61–67.

P. Dugac
[1] Fondements de l'analyse. In J. Dieudonné [1, vol. 1, chapter 6, pp. 335–392].

[2] *Histoire du théorème des accroissements finis*, Paris: Université Pierre et Marie Curie, 1979.

[3] *Limite, point d'accumulation, compact*, Paris: Université Pierre et Marie Curie, 1980.

[4] Elements d'analyse de Karl Weierstrass, *Archive for History of Exact Sciences* 10(1973): 41–176.

G. Eneström
[1] Über eine von Euler aufgestellte allgemeine Konvergenzbedingung, *Bibliotheca mathematica* (3) 6(1905): 186–189.

Euclid
[1] *The Thirteen Books of Euclid's Elements*, ed. T. L. Heath, Cambridge: 1908. Reprinted in 3 vols., New York: Dover, 1956.

L. Euler
[1] De progressionibus harmonices observationes, *Commentarii*

academiae scientiarum . . . petropolitanae 7(1740). In Euler [6, series, 1, vol. 14, pp. 87–100].

[2] De seriebus divergentibus, *Novii commentarii academiae scientiarum petropolitanae* 5(1760). In Euler [6, series 1, vol. 14, pp. 585–617].

[3] *Institutiones calculi differentialis*, St. Petersburg: 1755. In Euler [6, series 1, vol. 10].

[4] *Institutiones calculi integralis*, 3 vols., St Petersburg: 1768–1770. In Euler [6, series 1, vols. 11–13].

[5] *Introductio in analysin infinitorum*, Lausanne: Bousquet, 1748. In Euler [6, series 1, vols. 8–9].

[6] *Opera omnia*, 3 series, Leipzig, Berlin, and Zurich: Teubner, 1911–.

[7] Rémarques sur un beau rapport entre les séries des puissances tant direct que réciproques, *Mémoires de l'Académie des Sciences de Berlin* (1761). In Euler [6, series 1, vol. 15].

[8] *Vollstandige Anleitung zur Algebra*, St. Petersburg: 1770. Reprinted, ed. J. E. Hofmann, Stuttgart: Reclam-Verlag, 1959.

P. M. Flett

[1] Some historical notes and speculations concerning the mean-value theorem of the differential calculus, *Bulletin of the Institute of Mathematics and Its Applications* 10(1974):66–72.

J. Fourier

[1] Question d'analyse algébrique, *Bulletin des sciences par la Societé Philomathique* (1818):61–67. In Fourier [3, vol. 2, pp. 243–253].

[2] *Théorie analytique de la chaleur*, Paris: F. Didot, 1822. Reprinted in English translation: *Analytical Theory of Heat*, New York: Dover, 1955.

[3] *Oeuvres de Fourier*. 2 vols. Paris: Gauthier-Villars, 1888–1890.

H. Freudenthal

[1] Cauchy. In *Dictionary of Scientific Biography* vol. 3., New York: Scribners, 1971, pp. 131–148.

[2] Did Cauchy Plagiarize Bolzano?, *Archive for History of Exact Sciences* 7(1971):375–392.

P. Funk

[1] Bolzano als Mathematiker, *Sitzungs-Berichte Österreicher Akademie der Wissenschaften Wien* 252(1967) Part 5:121–134.

K. F. Gauss

[1] Disquisitio generales circa seriem infinitam

$$1 + \frac{\alpha \cdot \beta}{1 \cdot \gamma} x + \frac{\alpha(\alpha+1)\beta(\beta+1)}{1 \cdot 2 \cdot \gamma \cdot (\gamma+1)} x^2$$
$$+ \frac{\alpha(\alpha+1)(\alpha+2)\beta(\beta+1)(\beta+2)}{1 \cdot 2 \cdot 3 \cdot \gamma \cdot (\gamma+1)(\gamma+2)} x^3 + \cdots,$$

Commentationes societatis regiae scientiarum gottingensis (1813): 1–46. In Gauss [2, pp. 123–162].

[2] *Werke*, vol. 3, Gottingen: 1866. Reprinted Hildesheim: George Olms, 1973.

H. S. Gerdil

[1] De l'infini absolu, *Miscellanea taurinensia* 2(1760–1761): 1–45.

C. C. Gillispie

[1] *Lazare Carnot Savant*, Princeton: Princeton University Press, 1971.

H. Goldstine

[1] *A History of Numerical Analysis from the 16th through the 19th Century*, New York, Heidelberg, and Berlin: Springer, 1977.

J. V. Grabiner

[1] Cauchy and Bolzano: Tradition and transformation in the history of mathematics. In *Transformation and Tradition in the Sciences*, ed. E. Mendelsohn, forthcoming.

[2] Changing attitudes toward mathematical rigor: Lagrange and analysis in the eighteenth and nineteenth centuries. In *Epistemologische und soziale Probleme der Wissenschaftsentwicklung im frühen 19. Jahrhundert*, ed. M. Otte, Dordrecht: Reidel, to appear.

[3] Is mathematical truth time-dependent?, *American Mathematical Monthly* 81(1974): 354–365.

[4] The origins of Cauchy's theory of the derivative, *Historia mathematica* 5(1978): 379–409.

I. Grattan-Guinness

[1] Berkeley's criticism of the calculus as a study in the theory of limits, *Janus* 56(1969): 215–227.

[2] Bolzano, Cauchy and the "new analysis" of the early nineteenth century, *Archive for History of Exact Sciences* 6(1970): 372–400.

[3] *The Development of the Foundations of Mathematical Analysis from Euler to Riemann*, Cambridge, Massachusetts: MIT Press, 1970.

[4] Preliminary notes on the historical significance of quantification and of the axioms of choice in the development of mathematical analysis, *Historia mathematica* 2(1975): 475–488.

J. P. Gruson

[1] Le calcul d'exposition, *Mémoires de l'Académie des Sciences de Berlin* (1798): 151–216.

[2] Suite du mémoire sur le calcul d'exposition, *Mémoires de l'Académie des Sciences de Berlin* (1799–1800): 177–189.

A. R. Hall

[1] *The Scientific Revolution*, 2nd ed., Boston: Beacon Press, 1962.

R. Hamburg

[1] The theory of equations in the eighteenth century: The work of Joseph Lagrange, *Archive for History of Exact Sciences* 16(1977): 17–36.

T. Hankins

[1] *Jean d'Alembert*, Oxford: Oxford University Press, 1970.

T. Hawkins

[1] *Lebesgue's Theory of Integration: Its Origins and Development*, Madison: University of Wisconsin Press, 1970.

T. L. Heath

[1] *History of Greek Mathematics*, 2 vols., Oxford: Clarendon Press, 1921.

[2] ed., *The Works of Archimedes*, Cambridge: Cambridge University Press, 1912.

E. Heine

[1] Die Elemente der Funktionenlehre, *Crelles Journal* 74(1872): 172–188.

J. E. Hofmann

[1] *Geschichte der Mathematik*, 3 vols., Berlin: Walter de Gruyter, 1953–1957.

G. F. A. de l'Hôpital

[1] *Analyse des infiniment petits*, Paris: Imprimérie Royale, 1696.

S. L'Huilier

[1] *Exposition élémentaire des principes des calculs supérieurs*, Berlin: Decker, 1787.

[2] *Principiorum calculi differentialis et integralis*, Tubingen: Cottam, 1795.

C. Hutton

[1] *A Mathematical and Philosophical Dictionary*, London: J. Davis, 1795.

R. F. Iacobacci

[1] *Augustin-Louis Cauchy and the development of mathematical anal-*

ysis, Unpublished Doctoral Dissertation, New York University, 1965.

A. P. Iushkevich

[1] Euler, *Dictionary of Scientific Biography*, vol. 4, New York: Scribner's, 1971, pp. 467–484.

[2] Euler und Lagrange über die Grundlagen der Analysis, *Sammelband der zu ehren des 200 Geburtstages Leonhard Eulers*, Berlin: Deutschen Akademie der Wissenschaften zu Berlin, 1959, pp. 224–244.

[3] J. A. da Cunha et les fondements de l'analyse infinitésimale, *Revue d'histoire des sciences* 26(1973): 1–22.

[4] Lazare Carnot and the competition of the Berlin Academy in 1786 on the mathematical theory of the infinite. In C. C. Gillispie, *Lazare Carnot Savant*, Princeton: Princeton University Press, 1971.

[5] On the origins of Cauchy's concept of the definite integral [o vozniknoveniya poiyatiya ob opredelennom integrale Koshi: in Russian], *Trudy Instituta Istorii Estestvoznaniya, Akademia Nauk SSSR* 1(1974): 373–411.

[6] The concept of function up to the middle of the nineteenth century, *Archive for History of Exact Sciences* 16(1977): 37–85.

P. E. B. Jourdain

[1] The ideas of the "Fonctions analytiques" in Lagrange's early work, *Proceedings of the International Congress of Mathematicians* 2(1912): 540–541.

[2] The origins of Cauchy's conception of the definite integral and of the continuity of a function, *Isis* 1(1913): 661–703.

[3] The theory of functions with Cauchy and Gauss, *Bibliotheca mathematica* (3), 6(1905): 190–207.

P. Kitcher

[1] Bolzano's ideal of algebraic analysis, *Studies in History and Philosophy of Science* 6(1975): 229–269.

[2] Fluxions, limits, and infinite littlenesse: A study of Newton's presentation of the calculus, *Isis* 64(1973): 33–49.

F. Klein

[1] *Vorlesungen über die Entwicklung der Mathematik im 19ten Jahrhundert*, 2 vols., Berlin: 1926–1927. Reprinted New York: Chelsea, 1967.

M. Kline

[1] *Mathematical Thought from Ancient to Modern Times*, New York: Oxford University Press, 1972.

G. S. Klügel
[1] *Mathematisches Wörterbuch*, Leipzig: Schwickert, 1803–1831.

A. Kolman
[1] *Bernard Bolzano*, Berlin: Akademie Verlag, 1963.

E. Koppelman
[1] The calculus of operations and the rise of abstract algebra, *Archive for History of Exact Sciences* 8(1971): 155–242.

G. Kowalewski
[1] Über Bolzanos nichtdifferenzierbare stetige Funktion, *Acta mathematica* 44(1923): 315–319.

T. Kuhn
[1] *The Structure of Scientific Revolutions*, 2nd ed., Chicago: International Encyclopedia of Unified Science, 1970.

A. I. Kurdyumova
[1] The Bolzano–Cauchy convergence criterion in an 1812 work of Gauss [in Russian], *Istoriko-Matematicheskiye Issledovania* 23(1978): 142–143.

S. F. Lacroix
[1] *Traité du calcul différentiel et du calcul intégral*, 1st ed., 3 vols., Paris: Duprat, 1797.

[2] *Traité du calcul différentiel et du calcul intégral*, 2nd ed., 3 vols., Paris: Courcier, 1810–1819.

[3] *Traité élémentaire de calcul différentiel et de calcul intégral*, Paris: Duprat, 1802.

J. L. Lagrange
[1] *De la résolution des équations numériques de tous les degrés*, Paris: Duprat, An VI (1798).

[2] Discours sur l'object de la théorie des fonctions analytiques, *Journal de l'Ecole Polytechnique*, Cahier 6, 2(1799). In Lagrange [9, vol. 7, pp. 323–328].

[3] *Leçons élémentaires sur les mathématiques données à l'école normale en 1795*, Seances des Ecoles Normales (1794–1795). In Lagrange [9, vol. 7, pp. 181–288].

[4] *Leçons sur le calcul des fonctions*, new ed. Paris: Courcier, 1806. In Lagrange [9, vol. 10].

[5] Letter to d'Alembert, 15 July 1769. In Lagrange [9, vol. 13, pp. 137–143].

[6] *Mechanique analitique*, Paris: Desaint, 1788.

[7] *Mécanique analytique*, 2nd ed. 2 vols., Paris: Courcier, 1811–1815. In Lagrange [9, vols. 11, 12].

[8] Note sur la métaphysique du calcul infinitésimal, *Miscellanea taurinensia* 2(1760–1761):17–18. In Lagrange [9, vol. 7, pp. 597–599].

[9] *Oeuvres de Lagrange*, publiées par les soins de M. J.-A. Serret, 14 vols., Paris: Gauthier-Villars, 1867–1892. Reprinted Hildesheim and New York: Georg Oms Verlag, 1973.

[10] Et al, Prix proposés par l'Académie Royale des Sciences et Belles-Lettres pour l'année 1786, *Nouveaux mémoires de l'Académie Royale des Sciences et Belles-Lettres* (1784):12–14.

[11] Et al., Prix proposés par l'Academie Royale des Sciences et Belles-Lettres pour l'année 1788, *Mémoires de l'Académie Royale des Sciences et Belles-Lettres* (1786):8–9.

[12] Réflexions sur la résolution algébrique des équations, *Nouveaux mémoires ... Berlin* (1770):134–215; (1771):138–253. In Lagrange [9, vol. 3, pp. 205–424].

[13] Sur la résolution des équations numériques, et additions au mémoire sur la résolution des équations numériques, *Mémoires de l'Académie ... Berlin*, 23(1767):311–352; 24(1768):111–180. In Lagrange [9, vol. 2, pp. 539–654].

[14] Sur le calcul des fonctions, *Séances des Ecoles Normales* 10(1801).

[15] Sur une nouvelle espèce de calcul rélatif à la différentiation et à l'intégration des quantités variables, *Nouveaux mémoires ... Berlin*, Classe de mathématiques (1772):185–221. In Lagrange [9, vol. 3, pp. 439–476].

[16] *Théorie des fonctions analytiques*, Paris: Imprimérie de la République, An V (1797).

[17] *Théorie des fonctions analytiques*, new ed., Paris: Courcier, 1813. In Lagrange [9, vol. 9].

[18] *Traité de la résolution des équations numériques de tous les degrés*, 2nd ed., Paris: Courcier, 1808. In Lagrange [9, vol. 8].

I. Lakatos
[1] *Proofs and Refutations: The Logic of Mathematical Discovery*, ed. J. Worrall and E. Zukor, Cambridge: Cambridge University Press, 1976.

J. H. Lambert
[1] Observationes variae in mathesin puram, *Acta Helvetica* 3(1758):128–168.

J. Landen
[1] *The Residual Analysis*, London: for the Author, 1764.

P.-S. Laplace
[1] Leçons de mathématiques données à l'école normale, 1795, *Journal de l'Ecole Polytechnique* 7 : 1–172.

H. Lebesgue
[1] *Leçons sur l'intégration et la récherche des fonctions primitives*, Paris: Gauthier-Villars, 1904.

A. M. Legendre
[1] *Exercices de calcul intégral*, 3 vols., Paris: Courcier, 1811–1817.

G. W. Leibniz
[1] De geometria recondita et analysi indivisibilium atque infinitarum, *Acta eruditorum* 5(1686). In Leibniz [2, vol. 3, pp. 226–235]).

[2] *Mathematische Schriften*, ed. G. I. Gerhardt, 7 vols., vols. 1–2, Berlin: Asher; vols. 3–7, Halle: Schmidt, 1849–1863.

[3] Mémoire touchant son sentiment sur le calcul différentiel, *Journal de Trevoux* (1701) : 270–272. In Liebniz [2, vol. 1, p. 350].

[4] Nova methodus pro maxima et minima ..., *Acta eruditorum* 3(1684) : 467–473. In Leibniz [2, vol. 3].

C. Maclaurin
[1] *A Treatise of Algebra*, 2nd ed., London: A. Miller and J. Nourse, 1756.

[2] *A Treatise of Fluxions in Two Books*, Edinburgh: Ruddimans, 1742.

K. Manning
[1] The emergence of the Weierstrassian approach to complex analysis, *Archive for History of Exact Sciences* 14(1975) : 297–383.

K. O. May
[1] *Bibliography and Research Manual of the History of Mathematics*, Toronto and Buffalo: University of Toronto Press, 1973.

E. Mendelsohn
[1] The emergence of science as a profession in nineteenth-century Europe. In Karl Hill, ed., *The Management of Scientists*, Boston: Beacon Press, 1964, pp. 3–48.

C. Méray
[1] *Nouveau précis d'analyse infinitésimale*, Paris: F. Savy, 1872.

J. T. Merz
[1] *History of European Thought in the 19th Century*, 4 vols., London: Blackwell, 1904–1912. Reprinted New York: Dover, 1965.

G. G. Mittag-Leffler
[1] Die ersten 40 Jahre des Lebens von Weierstrass, *Acta mathematica* 39(1921) : 1–56.

F. N. M. Moigno
[1] *Leçons de calcul différentiel et de calcul intégral, redigées d'apres les méthodes et les ouvrages ... de M. A.-L. Cauchy*, 2 vols., Paris: Bachelier, 1840–1844.

[2] *Vorlesungen über die Integralrechnung, vorzüglich nach den Methoden von A. L. Cauchy bearbeitet*, Braunschweig: G. C. E. Meyersen, 1846 (German translation of Moigno [1]).

I. Newton
[1] *Mathematical Principles of Natural Philosophy* [1727], tr. A. Motte, rev. F. Cajori, Berkeley: University of California Press, 1934.

[2] *Mathematical Works of Isaac Newton*, 2 vols, ed. D. T. Whiteside, New York and London: Johnson Reprint, 1964–1967.

[3] *Method of Fluxions*, London: 1737 (written in Latin, 1671). In Newton [2, vol. 1. pp. 29–139].

[4] *Of Analysis by Equations of an Infinite Number of Terms*, London: J. Stewart, 1745 (written 1669). In Newton, [2, vol. 1, pp. 3–25].

[5] *Quadrature of Curves*, London: 1710 (written in Latin, 1693; published in Latin, 1704). In Newton [2, vol. 1, pp. 141–160].

[6] *Universal Arithmetic*, London: 1728 (written in Latin and published in Latin, 1707). In Newton [2, vol. 2, pp. 3–134].

O. Ore
[1] *Neils Henrik Abel*, Basel: Birkhauser, 1950.

J. L. Ovaert
[1] La thèse de Lagrange et la transformation de l'analyse. In J. L. Ovaert et al., *Philosophie et calcul de l'infini*, Paris: Maspero, 1976, pp. 157–200.

P. Painlevé
[1] Gewöhnliche Differentialgleichungen. Existenz der Lösungen, *Enzyclopädie der Mathematischen Wissenschaften* IIB1, 1, Section II, A4a, pp. 189–293.

J. Pierpont
[1] Mathematical rigor, past and present, *Bulletin of the American Mathematical Society* 34(1928):23–53.

H. Poincaré
[1] *Compte rendu du 2me congrés internationale des mathématiciens, 1900*, Paris: 1902, pp. 120–122.

S.-D. Poisson
[1] Suite du mémoire sur les intégrales définies, *Journal de l'Ecole Polytechnique*, Cahier 18, 11(1820):295–341.

A. Pringsheim and J. Molk

[1] Principes fondamentaux de la théorie des fonctions. In *Encyclopédie de sciences mathématiques pures et appliquées* 2, vol. 1, fascicle 1, 1909.

R. Reiff

[1] *Geschichte der unendlichen Reihen*, Tubingen: H. Laupp, 1889.

B. Riemann

[1] *Gesammelte mathematische Werke*, 2nd ed., ed. H. Weber, Leipzig: 1892. Reprinted New York: Dover 1953; includes "Nachtrage," originally published 1902.

[2] Ueber die Darstellbarkeit einer Function durch eine trigonometrische Reihe, *Abhandlungen der Königliche Gesellschaft der Wissenschaft Gottingen* 13 (1867). In Riemann [1, pp. 227–265].

B. Robins

[1] *Mathematical Tracts*, 2 vols., London: J. Nourse, 1761.

A. Robinson

[1] *Non-Standard Analysis*, Amsterdam: North-Holland Press, 1970.

B. van Rootselaar

[1] Bolzano, *Dictionary of Scientific Biography*, vol. 2, New York: Scribner's, 1970, pp. 273–279.

S. B. Russ

[1] A translation of Bolzano's paper on the intermediate value theorem, *Historia mathematica* 7(1980): 156–185.

K. Rychlik

[1] *Theorie der reelen Zahlen in Bolzanos handschriften Nachlasse*, Prague: Akademie der Wissenschaften, 1962.

G. Sarton

[1] *The Study of the History of Mathematics*, Cambridge, Massachusetts: Harvard Press, 1936. Reprinted New York: Dover, 1957.

H. Sinaceur

[1] Cauchy et Bolzano, *Revue d'histoire des sciences* 26(1973): 97–112.

P. Staeckel

[1] Integration durch das imaginare Gebeit, *Biblioteca mathematica* (3) 1(1900): 109–128.

O. Stolz

[1] B. Bolzanos Bedeutung in der Geschichte der Infinitesimalrechnung, *Mathematische Annalen* 18(1881): 255–279.

D. J. Struik
[1] Ed., *A Source Book in Mathematics, 1200–1800*, Cambridge, Massachusetts: Harvard University Press, 1969.

[2] *Concise History of Mathematics*, New York: Dover, 1967.

L. Sylow
[1] Les études d'Abel et ses découvertes. In *Neils Henrik Abel: Mémorial publié à l'occasion du centenaire de sa naissance*, Kristiania: J. Dybwad, 1902.

C. Truesdell
[1] The rational mechanics of flexible or elastic bodies, 1638–1788. In Euler [6, series 2, vol. 11, part 2].

C. A. Valson
[1] *La vie et les travaux du baron Cauchy*, Paris: Gauthier-Villars, 1868.

G. Vivanti
[1] Infinitesimalrechnung. In M. Cantor [1, vol. 4, pp. 639–869].

A. Voss
[1] Differential- und Integralrechnung. In *Enzyclopädie der mathematischen Wissenschaften*, II A II, Leipzig: Teubner, 1899, pp. 54–134.

B. L. van der Waerden
[1] *Science Awakening*, New York: Wiley, 1963.

K. Weierstrass
[1] *Mathematische Werke*, 7 vols., Berlin: 1894–1915. Reprinted Hildesheim and New York: Olms, 1967.

E. Winter
[1] *Bernhard Bolzano: Ein Denker und Erzieher im Oesterreicheschen Vormärz*, Vienna: Böhlaus, 1967.

K. Zimmermann
[1] Arbogast als Mathematiker und Historiker der Mathematik, Dissertation, Heidelberg: 1934.

Index